Acting with the World

Acting with the World

AGENCY IN THE ANTHROPOCENE

World

....

Andrew Pickering

DUKE UNIVERSITY PRESS   DURHAM AND LONDON   2025

Project Editor: Bird Williams
Designed by Matt Tauch
Typeset in MeropeBasic, Work Sans, and Comma Base
by Copperline Book Services.

Library of Congress Cataloging-in-Publication Data
Names: Pickering, Andrew, [date] author.
Title: Acting with the world : agency in the Anthropocene /
Andrew Pickering.
Description: Durham : Duke University Press, 2025. | Includes
bibliographical references and index.
Identifiers: LCCN 2024026740 (print)
LCCN 2024026741 (ebook)
ISBN 9781478031512 (paperback)
ISBN 9781478028307 (hardcover)
ISBN 9781478060499 (ebook)
Subjects: LCSH: Nature—Effect of human beings on. | Human
ecology. | Geology, Stratigraphic—Anthropocene. | Global
environmental change—Social aspects.
Classification: LCC GF75 . P525 2025 (print)
LCC GF75 (ebook)
DDC 304.2—dc23/eng/20241015
LC record available at https://lccn.loc.gov/2024026740
LC ebook record available at https://lccn.loc.gov/2024026741

Cover art: Cy Gavin, *Untitled (Gibbet Island)*, 2019. Acrylic,
pink Bermuda sand, aquarelle pencil, chalk and oil on
canvas, 92 × 92 × 1¾ in. (233.7 × 233.7 × 4.5 cm). Purchase,
gift of Mr. and Mrs. Richard M. Scaife, by exchange, 2019.54.
Carnegie Museum of Art, Pittsburgh, PA/Art Resource, NY. ©
Cy Gavin. Courtesy of the artist and Carnegie Museum of Art.

Contents

As individual animals we are not so special, and in some ways the human species is like a planetary disease. —JAMES LOVELOCK, *The Revenge of Gaia*

What shall we do? No one yet knows. Unless we think about fundamentals, our specific measures may produce new backlashes more serious than those they are designed to remedy. —LYNN WHITE JR., "The Historical Roots of Our Ecological Crisis"

A new paradigm would have to take up practices that are now on the margin of our culture and make them central. —HUBERT DREYFUS, "Heidegger on the Connection between Nihilism, Art, Technology, and Politics"

I would like to cultivate a charisma of uncertainty, a charisma of admitting that you're making it up as you go along.... I think we're in for a hard ride for maybe half a century. Then it will either be the end of civilization or a reborn humanity with a different set of ideas about who we are and where we belong and how we must relate to things in order to survive. —BRIAN ENO, quoted in David Marchese, "Brian Eno Reveals the Hidden Purpose of All Art"

Living with and dying with each other potently in the Chthulucene can be a fierce reply to the dictates of both Anthropos and Capital. —DONNA HARAWAY, *Staying with the Trouble*

Preface

All my subsequent work has grown out of a book I published in 1995, *The Mangle of Practice: Time, Agency, and Science,* in which I drew on case studies of scientific research to develop a general story of what the world is like, a worldview, an ontology. What has fascinated me ever since is that the worldview I spelled out there was different from that of the physicists I had been studying. They took it for granted that the world is built from fixed and knowable entities like quarks or strings or black holes, while the world I found myself describing in *The Mangle* was fluid, always evolving and becoming something new.

At first I did not think much about that. The scientists had their worldview, I had mine, and that was OK. But then I started to take the divergence more seriously. I began to think we were both right. I was right to say the world is a place of open-ended becoming — you just have to look to see that — and that the scientists themselves indeed live in that world. But at the same time, they imposed on their work the objective of discovering more-or-less stable findings in the flux — islands of stability, as I later called them (Pickering 2017b). Only more-or-less stable facts, instruments, and machines are allowed to count as the products of science: Newton's laws, electrons, bubble chambers. To find any such islands is a heroic achievement, but the price is to conceal all the evolution and becoming that I felt we need to talk about — in order, even, to understand how science itself works.

So what? One can, I suppose, be interested in ontology as a topic in its own right, but at this point I started to wonder what sort of implications different ontologies have for *practice,* for how we *act* in the world. It seemed to me that modern sciences like physics and chemistry are, so to speak, predicated on a dualist ontology that makes a clean split between people and things, in the sense that maintaining this split is the criterion of successful science. Would we perhaps go on differently if we assumed instead an ontology of becoming?

At first I was baffled by this question, but in the late 1990s, I came across an odd and almost forgotten science (if it is a science) called cybernetics, which shared my view of the world as an ultimately unknowable space of open-ended transformation. And, crucially, the cyberneticians put this worldview into practice — they brought it to life and acted it out in all sorts of projects in all sorts of fields running from brain science, psychiatry, and robotics to management, the arts, even spirituality. These projects showed me what I had been unable to imagine, what a mangle-ish ontology could look like in practice. I found them fascinating in their difference, their strangeness, their departure from their conventional equivalents, and their imaginative quality — I could never have made them up. I looked, for example, at adaptive architecture, buildings designed to become something new, to change shape in response to how they were used, and antipsychiatry, an approach in which the psychiatrists lived communally with the mad in processes of reciprocal adaptation, rather than prescribing them drugs or shock therapy.

I felt that I had turned up a new paradigm, a new world — a new way of grasping and acting in the world — very different from the usual ways of doing things that I was familiar with, all interlinked via the ontology of becoming I had developed earlier. And I was drawn to this paradigm. I liked the idea of buildings that changed shape; it would be interesting finding out where the kitchen had gone when I woke up in the morning. Antipsychiatry was controversial, but I admired men and women who would try to help very disturbed people by living with them and transforming themselves in the process. So I wrote a book, *The Cybernetic Brain: Sketches of Another Future* (2010), tracing out the main lines of development of cybernetics since its first appearance in the 1940s, especially as it had evolved in Britain.

Since then, much of my work has focused on tracing out further the contours of a neo-cybernetic paradigm, without worrying too much whether the projects I examined called themselves cybernetic or not. I have followed two leads in particular. One concerned unconventional cybernetic artworks that somehow show us that we live in a lively world of endless becoming, as a kind of ontological pedagogy (Pickering, forthcoming a). The other is the topic of this book and concerns our relations with nature and the environment.

My inspiration here was Gregory Bateson, one of the first generation of cyberneticians. He featured in *Cybernetic Brain* by virtue of his connection to antipsychiatry (and Buddhism), but I also knew that later in his life, in the late 1960s, Bateson was part of the environmental movement in the United

States (G. Bateson 1968, 2000b; M. Bateson 2005). He felt that the environmental crises of the time, crystallized in Rachel Carson's *Silent Spring* (1962), were just surface symptoms of a deeper malaise in our relations with nature and that we needed to engage with nature differently.

Bateson's main concern was with ontology, arguing against the dualist worldview and in favor of a recognition of couplings of the human and nonhuman worlds — "the pattern that connects," as he put it (G. Bateson [1979] 2002). Unfortunately, from my perspective, he offered few suggestions for novel forms of nondualist practice. But his work encouraged me to look for neo-cybernetic ways of getting closer to nature, acting with rather than on the environment, ways that would sensitize us to the world we are irrevocably plunged into, rather than, as Bateson feared, cutting us off from it. As before, I started coming across examples of this new paradigm, now in our relations with nature, and the present book is about what I found. At its heart are a series of studies set out in the following chapters, which touch on earth, fire, and water (though not much on air) and, in the end, spirits.

I can think of several reasons for being interested in these. (1) They are simply and systematically different from our usual patterns of action. I find them surprising and new in ways that are worth contemplating. (2) They show us that we have a choice. Our usual ways of going on are not dictated by the order of things; we can therefore act differently if we want to. (3) The environmental crises that worried Bateson have only gotten worse since the 1960s, with global warming as the poster child. A choice of acting differently — and less perilously — is thus more valuable than ever, though I should say now that I have no quick fix to offer for the ills of the Anthropocene. (4) They bring us closer to nature, reminding us of our inseparable coupling to it and even rejoicing in that, attuning us to its ways. In the end, that might be what we need most.

....

I began the first draft of this book during the coronavirus years — a solitary affair — but before that much of the thinking grew out of my undergraduate and graduate seminars at the University of Illinois at Urbana-Champaign and the University of Exeter, and I thank the students who took part in them. They will recognize some of the studies that follow, but I have taken them further and hope to have made more sense of them. For input, feedback, conversation, and enlightenment, I want also to express my gratitude for individual contributions from Lisa Asplen, Antonio Carvalho, Dawn Coppin,

Giovanna Columbetti, John Dupré, Adrian Franklin, Regenia Gagnier, Steve Hinchliffe, Casper Bruun Jensen, Pablo Jensen, Bruce Lambert, Lenny Moss, Paul Pangaro, Simon Penny, Brian Rappert, James Rice, Chris Salter, Ernesto Schwartz-Marin, Tom Smith, and Chris Welsby. In 2022, Steve Hinchliffe organized a group discussion of the first draft of this book which fed importantly into the present version, and I am very grateful to Steve and the participants for that. At Duke University Press I thank Ken Wissoker, Ryan Kendall, Bird Williams, and two anonymous readers. Lastly, my thanks go to Lenny Moss and Paul O'Connor for the gift of office space on campus — greatly appreciated!

Introduction

Acting on or with the World

The topic of this book is an unfamiliar pattern of action that I call *acting-with* or *poiesis*, a pattern that entails paying attention to the tendencies of the world, incorporating them into our ways of going on, and tuning our own activities into them. The goal of the following chapters is to exemplify acting-with in our relations with nature and the environment — to bring poiesis to life in the time of the Anthropocene. Along the way and to emphasize the distinctive aspects of acting-with, I find it necessary to discuss a contrasting and much more recognizable pattern of action that I call *acting-on* or *enframing*, but I should say now that my goal is *not* a balanced and comprehensive portrayal or critique of acting-on; my central concern throughout is acting-with.

Acting on the world is the very familiar stance of mastery and domination that is the hallmark of modernity and which has brought us both modern science and the vast array of machines and technologies that underwrite the knowledge and power of the modern West. It is not to be taken lightly. But it is increasingly recognized that acting-on also has a dark side. In the Anthropocene, technological disasters, pollution, extinction of species, floods, fires, and global warming appear as corollaries of mastery and hubris. We are great at converting fossil fuels to mechanical power and electrical energy, for example, but we are also great at simultaneously producing carbon dioxide and climate change as an unintended and unwanted spin-off. Many people now feel we have gone too far along this trajectory and that it needs to be resisted, stopped, blocked in its tracks — Extinction Rebellion and Just Stop Oil being the most visible recent manifestations of this.

I share these sentiments, but I am no expert on resistance and I have nothing new to say about it. Nor am I concerned here with technological fixes from geo-engineering to solar power. I see them as still part of the acting on, dominating paradigm (though now in a therapeutic spirit). In contrast, I want to open up a different and less familiar space. I want to look at a level below that of domination, resistance, and technological fixes. I want to examine another pattern of acting in the world, acting-with. I want to signal the possibility of *systematically different ways of acting in the world*. My strategy in the following chapters is to describe and analyze a series of examples of acting with nature and the environment as a blueprint for the future. For the remainder of this chapter, I set out the basic perspective.

. . . .

What men want to learn from nature is how to use it in order wholly to dominate it and other men. —MAX HORKHEIMER AND THEODOR W. ADORNO, *Dialectic of Enlightenment*

"Acting with" is like "being nice" — it's hard to be against it but it doesn't mean much in itself. To put some flesh on the idea, we can start with the philosophical position known as dualism. Running from the Greeks through René Descartes to the present, this is the worldview that is central to Western modernity (Latour 1993). For our purposes, the duality in question is of people and things, the human and the nonhuman, understood to be different in kind. And this difference is usually understood asymmetrically, hierarchically. We humans have something special and exceptional — souls, reason, will — that sets us not just apart from the rest of creation but also above it, so to speak, in control. Dualism casts us as the only genuine agents in a passive and subservient world.

Dualism is what we implicitly teach our children in schools and universities when we teach them separately about the natural sciences (things) and the humanities and social sciences (people). And our made world echoes dualism back to us, filled with machines like cars and computers that usually obey our commands, slaves of their human masters. Dualism is our natural ontological attitude, one could say. And while dualism does not logically imply a stance of domination, the two fit together nicely: If we are the only genuine agents in the world, what else should we do but act on it and order it to serve us?

So, I want to say that domination as a pattern of *acting on* things, and indeed people, is a *dualist* way of going on. And the organizing question for this book is: what could a nondual way of going on look like? And how can we recognize it and emulate it when we see it?

. . . .

As a self-conscious philosophy, dualism went out of fashion long ago. It might be the natural ontological attitude of contemporary Western culture, but few academic philosophers would defend it today. And thinking about what is wrong with it will get us to *acting-with*. In the most general terms, what is wrong is simply that it is mistaken. Humanity is not, in fact, dualistically split off from the world. We are part of it. And many paths diverge at this point around the question of just how this "part of" is to be construed.[1] Here I will focus on my own way forward. In *The Mangle of Practice*, I developed a broadly pragmatist position which focused on *agency* — doing things, performance in the world — as constitutive of our being, and this emphasis on action and performance (rather than cognition) is central to my approach.[2] It brings out, first, a performative symmetry between the human and the nonhuman. We are certainly agents, we do consequential things in the world. But so are dogs and cats, stars and stones. And, second, beyond symmetry, the emphasis on agency brings out constitutive couplings across the dualist divide. The world responds to what we do to it and vice versa in a mutually transformative back-and-forth that I call, in a self-explanatory way, a *dance of agency* (Pickering 1995a).

Performance, agency, and the dance of agency are key concepts in all that follows. As just explained, they help map out a symmetric, nondualist worldview or ontology which, at the same time, denaturalizes domination and makes it problematic. A lively world does not need us to dominate and direct it and is quite capable of its own surprising performances (e.g., the recent coronavirus pandemic).

It is quite possible, then, to think our way out of dualism, but more needs to be said about patterns of action and about acting with, instead of on, the world. The first point to note is that while we are not, in fact, in charge of back-and-forth dances of agency with nature, we can act as if we are. In chapter 3, for example, we examine the attempts by the US Army Corps of Engineers (ACE) to dominate the Mississippi River and control its behavior. Of course, as we will also see, dances of agency never go away, but we could say that the ACE's actions veil or *background* the agency of the river. However the river replies to the ACE's initiatives, the ACE just plows on, trying to dictate terms to the river. In brief, then, this is how I want to think about dualist patterns of action: as backgrounding nonhuman agency, trying to ignore or suppress it, attempting to make the world dual. This is how I understand "acting on" the world.

And then, of course, a nondualist stance must entail something like *foregrounding* nonhuman agency, actively paying attention to what comes back to us from the world and seeing how we might get along with that. In chapter 4, for example, we see how dam operators have tried to learn from the effects of artificial floods to restore the ecosystem of the Grand Canyon — going with the flow, nonduality in action. This is how I understand acting-with.

Where have we got to? I began from a concern with domination and alternatives to it. And now we have an idea of the contrast between a dualist stance of domination as our usual pattern of action and a nondualist stance of acting-with as a systematic alternative to that, in terms of whether we somehow background or foreground the agency of nature. What remains is to bring these abstract formulations to life by examining some examples, which is what the rest of the book does. I review a series of studies of nondualist action in our relations with nature and the environment, trying to emphasize how different they are from our customary ways of going on. The object in the end is to open up an awareness of the possibilities for nondualist acting-with across the board. By way of contrast, as I said, I also discuss, relatively briefly, parallel forms of dualism in action in order to emphasize the specificity of acting-with.[3]

The following chapters run through my examples; but before that, various general observations are worth putting on the table.

....

1. *Words.* I am going to refer repeatedly to the two patterns of action just discussed — acting on and acting with — and it will help to have some more substantive adjectives and nouns to specify and distinguish them. I have thought about this a lot without arriving at any decisive solution. Dualist and nondualist is an obvious pairing, though not terribly evocative. Modern and nonmodern works (following Latour 1993). Human-centric or humanist versus decentered, posthuman. Martin Heidegger ([1954] 1977) called the dualist stance "enframing" — treating the world as "standing reserve" for human projects. His term for the other stance is "revealing," in the sense of "finding out" (e.g., finding out how a dance of agency will unfold) or "poiesis," the Greek word for "making" as "bringing forth." I will use all these words as appropriate but, carrying on from *The Cybernetic Brain*, mainly *enframing* for the stance of domination and *poiesis* for the nondualist alternative (pronounced, perhaps mistakenly, po-esis, with *poetic* as the adjective).[4] These terms seem richer and more substantive to me — the words themselves conjure up some of what is at stake in the different patterns of action that "dualist," "nonmodern," and so on do not. I know that *poiesis* is a strange word, but that is per-

haps appropriate for unfamiliar ways of acting. And to go back to the start, another way to catch the contrast in question, in a phrase rather than a word, is to think of enframing as *acting on* nature and poiesis as *acting with* nature.

2. *Paradigms.* Thomas Kuhn (1962) conceptualized major discontinuities in the history of science — scientific revolutions — as *gestalt switches*, in which aspects of the world that used to be in the foreground recede into the background and vice versa, bringing out new patterns. In just this sense, enframing and poiesis are different gestalts, different paradigms of thought and action, respectively backgrounding or foregrounding nonhuman agency. Kuhn argued that scientists act as if they live in different worlds before and after revolutions, and much the same is true of our examples of enframing and poiesis: poetic practices are strikingly different from their enframing counterparts.

3. *Emergence.* The poetic paradigm is decentered and posthuman, in the sense that poiesis evolves in the interplay of the human and the nonhuman with neither controlling the show. It is worth also emphasizing that this evolution is open-ended and *emergent*. The world can always surprise us. No one knows in advance how it will respond to our actions or vice versa. Dances of agency take shape in real time. Poiesis is thus an active and experimental process of *finding out* what will happen next. Again we can think here in terms of gestalts. Poiesis foregrounds emergence (as an integral aspect of agency), while enframing backgrounds it.

4. *Technology.* There is an important complication concerning both enframing and poiesis that needs attention. I have so far discussed both as *stances* in the world, dispositions to act in specific ways. The stance of enframing, say, means acting as if you are in charge, whatever comes along. The ACE's actions assume that the engineers can control the Mississippi even if they never quite succeed. But much of this book is concerned not so much with human action but with *technology* as a key interface with the world — so how should we think about that? I want to note that technologies, as well as stances, can be described as enframing or poetic, again precisely in the sense that they background or foreground the agency of their environments. In this respect, technologies are material proxies for human stances in the world.

Thus, most of the technologies that come immediately to mind — tools, machines, instruments — are technologies of enframing, designed to act on the world, to be *indifferent* to their surroundings, and *not* to engage in dances of agency. My laptop does not care that today is the hottest day this year in Britain; it acts just the same as if I were sitting in the snow. It would be useless to me if its word-processing actions varied with the weather or the time of day. Heidegger's ([1954] 1977) most memorable example of a technology is a hydroelectric plant straddling the Rhine. The plant does one thing and one thing only, turning water flow into electricity, whatever the river does. It pins the river down, so to speak, making it appear as a power source. In later chapters we will need to think about levees, dams, sea walls, and the like — "hard defenses" against the environment. All these are designed to be indifferent to their surroundings, technologies of enframing that dominate water without responding to it at all. We could say that technologies of enframing are things we impose on the world as fixed solutions, ways of gaining our specified ends. If you want to control a river and generate electricity, build a dam.[5]

As far as poetic technologies are concerned, two somewhat different categories will show up below. First, at the opposite pole from enframing must be no technology. We will encounter several examples of giving up the fight with nature and *letting go*, including the removal of enframing technologies, as a positive and constructive strategy to let the agency of nature shine through — a response to iatrogenic problems called up by prior interventions.

But there is a more intricate class of poetic technologies also to be found in our examples, technologies that *respond* to their worlds by engaging with them in dances of agency — acting with them. My earlier book, *The Cybernetic Brain*, is about machines and devices that do just that. But in the present book it is often better to widen the sense of "technology" to something like "techniques" or even just "practices." The examples of poiesis in later chapters are largely techniques and practices that are coupled into and act with their objects in regularized *choreographies of agency*, such as patterns of dam operation geared to river flows that serve to stabilize a downstream ecosystem (chapter 4). These interest me a lot.

5. *Knowledge and action.*

Science and technology are blessed words in our contemporary vocabulary.
—LYNN WHITE JR., "The Historical Roots of Our Ecological Crisis"

"I would have thought the science was sacred to you."
"The science is of course sacred."
—EMILY MAITLIS AND BERNARD-HENRI LÉVY, *Newsnight*. BBC 2

The future of Britain will depend on a new age of invention and innovation. Technological superpowers such as the United States and China are investing heavily.... Britain must find its niche in this new world. To do so requires a radical new policy agenda, with science and technology at its core, that transcends the fray of 20th-century political ideology. —SIR TONY BLAIR AND WILLIAM HAGUE, Baron Hague of Richmond, *A New National Purpose: Innovation Can Power the Future of Britain*

What's so great about science? —PAUL FEYERABEND, *Science in a Free Society*

So far I have discussed poiesis and enframing in performative terms, as patterns of action with or on the world. Now we need to think about knowledge. In *The Mangle* I argued for what I called a performative epistemology, a view of knowledge as entangled with and transformed in worldly practices and performances, and we can consider that further here. At the most basic level, some sort of everyday, commonsense knowledge is entailed in all the examples to follow, functioning as a way of keeping track of dances of agency in poiesis, for example — this happened, then that happened, and so on. Here I am more concerned with more organized forms of knowledge, especially the different roles of science in enframing and poiesis.

A theme that runs through what follows is that, as Heidegger ([1954] 1977) argued, in many ways the sciences are complicit in and even integral to enframing. They set the world up for enframing in at least two senses. At the most basic level, sciences like physics and chemistry conjure up and describe a dualist cosmos in which things obey laws and regularities quite independently of us. As discussed already, this dualism feeds directly into enframing. More specifically, the sciences show us the levers of power — if we do *this*, the world will do *that* — necessary for us to achieve our ends. And what interests me most in this connection is that, in contrast, science is largely *absent* from our examples of poiesis. I will qualify this in a moment, but for now my point is that poesis is, in this sense, *doing without science.*[6]

One way to see this is to think about emergence. I said before that enframing backgrounds emergence while poiesis foregrounds it, and we should note that science can be key to this backgrounding. If we can *calculate* how some system (say, a river) will respond to our actions (building a dam) we don't need to struggle poetically through any dances of agency, we don't need to find anything out. In enframing, science thus functions as a shortcut to the future, a way of knowing in advance what will happen, a detour around emergence; while poiesis foregrounds not knowledge but a performative finding-out in practice.[7]

This version of the contrast between acting-with and acting-on, poiesis and enframing, is worth paying attention to. It helps foreground the strangeness and unfamiliarity of poiesis. We routinely think of science and technoscience more broadly as the key to our future, in terms of increasing productivity and even addressing all the problems of the Anthropocene, but that leaves us in the space of enframing and makes poiesis very difficult to recognize and think about. Poetic possibilities get obscured by the focus on science, and one aim of this book is therefore to bring them out of the shadows.

I should emphasize that my interest in this ascientific aspect of poiesis is not an argument that we should abandon or get rid of science. But writing this book has made me see the place of science in the world differently. On the one hand, I want to note that science is central to and bound up with enframing and that, on the other, more importantly here, there exists another pattern, poiesis, to which science is, as it happens, less important. We will see in the studies to follow that poiesis can be a successful pattern of, literally, doing (acting, performing) without (any appeal to) science. That this other ascientific and nonmodern paradigm can exist and be taken seriously here and now in the early days of the third millennium is perhaps the most striking idea of this book.

One way to bring home the significance of these observations is to think about education. We know very well how to teach our children science and the so-called STEM (Science, Technology, Engineering, and Mathematics) subjects as key elements of the enframing paradigm, but there is little if any space in the curriculum for teaching the key performative aspects of poiesis and acting-with. We could almost say that modern education is an indoctrination into enframing and leaves poiesis unimaginable. This is important and we can return to it in the final chapter.

Now I need to qualify these remarks in two ways. First, while science is indeed absent from most of my examples of poiesis, there is a significant scientific aspect to one of them, the adaptive management of the Colorado River (chapter 4). We can explore that further when we get there, but I can note now that in that instance, science appears in a different guise from the way we usually think about it. Instead of a definitive guide to the future and a shortcut around emergence, science appears there in a more modest role, as an aid to "seeing in the dark," so to speak, and a warning of unknowability —a tentative and revisable guide to what might happen if we act this way or that, and a way, again, of keeping track of what has happened. This connects back to my interest in cybernetics and complexity science, to which we will later return.

Second, in a place analogous to science, various forms of Indigenous knowledge appear in chapters 7 and 8, on fire and spirits, respectively. I will not go into them in any detail, but I can mention here their nondualist aspect. Unlike modern scientific knowledge, they all foreground the agency of nature, they are guides to acting with the world rather than acting on it. They thus belong to the poetic rather than the enframing paradigm.

6. *Ontology*. The forms of knowledge just discussed are actively geared into practice. Both modern science and Indigenous knowledge help us see the future in courses of worldly action. But we could also think here about ontologies, overall visions of what the world is like, that implicitly inform and illuminate different patterns of practice, that are themselves proper to the enframing and poetic paradigms. Thus, I have already discussed how a dualist ontology which makes a clean and asymmetric split between people and things feeds into enframing as action (and vice versa). Likewise, the nondualist ontology I argued for in *The Mangle* hangs together with poiesis.

But my examples of poiesis invite some further ontological considerations. On the one hand, as I said before, cybernetics as a nonmodern science shares my mangle-ish ontology, and I find it interesting to explore ways in which specific resources from cybernetics illuminate the case studies of managing the Colorado River (chapter 4) and natural farming (chapter 6). Cybernetics does not help in seeing the future like the modern sciences; instead, it helps us get the hang of what is going on in maneuvers in fields of human and nonhuman agency and something of the strangeness of those maneuvers from a modern perspective.

In different ways, the studies also speak to other nonmodern ontological visions. In chapter 6, I argue that traditional Chinese concepts, specifically *shi* and *wu wei*, illuminate key aspects of natural farming (and thus our other studies, too), and that, conversely, the example of natural farming helps us grasp this unfamiliar non-Western ontology. Elsewhere, especially in the chapters on fire (chapter 7) and spirits (chapter 8), Native Australian and Amazonian animism appears, connecting back to the forms of Indigenous knowledge just mentioned. Again, animism is interesting in the present context as an ontology that foregrounds the agency of nature and our engagement with it. And if nothing else, animism, with its chancy gods and spirits, offers a warning about the possibility of technological failure.

7. So what?

As a source of life nature was venerated as sacred and human evolution was measured in terms of man's capacity to merge with her rhythms and patterns. —VANDANA SHIVA, *Staying Alive: Women, Ecology and Survival in India*

The fully enlightened earth radiates disaster triumphant. —MAX HORKHEIMER AND THEODOR W. ADORNO, *Dialectic of Enlightenment*

Enframing has brought us the immense power of modern technoscience. It *is* that immense power. So why bother ourselves about a different pattern of action, poiesis? For me, it began with a concern for the "political" implications of my ideas about agency and the mangle of practice — what sort of patterns of action do they point to? I find this question straightforwardly interesting, and the examples of poiesis that follow are the germ of an answer. From another angle, I have been using *poetic* as the adjectival form of *poiesis*, but it seems to me that there is also something poetic in an everyday sense and even graceful in the examples of poiesis in the following chapters. The opposite of *poetic* in this usage is something like *graceless*.[8] There is something graceless in drilling for oil a mile below the sea and hoping you can get away with it (Deepwater Horizon), as there is about enframing, mastery, slavery, and domination in general.

We can also think here about the perils of enframing — its dark side — and the corresponding promise of poiesis. Heidegger ([1954] 1977) described enframing as a "supreme danger" to humanity. On the one hand, he thought that enframing as a stance of mastery and domination distances and cuts us off from the world — nothing comes back from the slave to the master — and starves our inner being. We shrivel up inside.[9] Conversely, in foregrounding the nonhuman agency that enframing forgets, poiesis puts us in touch with the surprising powers of the world we live in and fosters intimate, sensitive, and responsive connections to it. This is the sense in which poiesis gets us closer to nature — performatively rather romantically.[10]

More concretely, in the wake of World War II, it was not hard for Heidegger to see technologies like gas chambers and atom bombs as horrifying exercises in enframing gone mad, to be avoided in the future at all costs. Seventy years or so later, as we travel deeper into the Anthropocene, the dark side of enframing is clearer. Emergence always bursts through somewhere. We have bent more and more of the world to our will but, as I said before, at the price of more and more — and bigger and bigger — unintended and unwanted "side effects": pollution, environmental disasters, global warming, the mass extinction of species.[11] Enframing looks more brutal and dangerous every day, and this is certainly a reason to be interested in graceful and poetic alternatives. Poiesis is no magic bullet, but we can see in the following chapters how poetic approaches can help obviate or even avoid the dangers of enframing, and thus chip away at the Anthropocene from below — it offers another way to be.

To come at the *So what?* question more positively, I can note that there is general agreement that many real-world problems today concern so-called wicked systems, meaning systems that somehow resist scientific analysis, that science just bounces off, for some reason (Rittel and Webber 1973). The ecosystem of the Colorado River would be a nice example — too immensely complex for the effects of our interventions to be meaningfully calculated even if we knew all the equations, which we do not. Enframing is thus at most a blunt and rather dangerous instrument for tackling wicked systems and the wicked problems that surround us. And poiesis, in contrast, indeed offers a constructive approach when science fails — finding out about the world performatively, not cognitively, feeling our way forward in experimental dances of agency, as in the adaptive management of the Colorado ecosystem (chapter 4). It strikes me that this a very significant reason for being interested in poetic approaches across a very broad front.[12]

8. *Going back.*

These developments [climatic changes] are making it ever more evident that many "savage" and "brutish" people understood something about landscapes and the Earth that their conquerors did not. This, perhaps, is why even hardheaded, empirically minded foresters, water experts, and landscape engineers have begun to advocate policies based on Indigenous understandings of ecosystems. —AMITAV GHOSH, *The Nutmeg's Curse*

Far from an intransigent attachment to the past, ancestrality stems from a living memory that orients itself to the ability to envision a different future. —ARTURO ESCOBAR, "Sustaining the Pluriverse"

While writing this book, it has dawned on me that many of my examples in one way or another involve going back in time, restoring what has been lost or finding another way we used to do things. Initially, I found this surprising and almost regrettable, but in retrospect it seems obvious. If our dominant pattern of action today is dualistic mastery, one obvious source of inspiration must be in the nondualist past — nothing comes from nowhere. So here

I just want to emphasize that this was, for me, a discovery. I did not begin this book as a nostalgic trip down memory lane or from a conviction that everything used to be better.[13] Instead, I have found myself interested in specific patterns of action, some of which are rooted in the past, but which are practical and manifested in the present as seeds for the future. My conclusion is that — sometimes, and in specific ways — we need to rewind history in order to reopen paths not taken.[14]

Something similar can be said with respect to the nonmodern ontologies mentioned earlier. I discuss these nonmodern philosophies and spiritualities as appropriate because I am struck by the ways in which they speak directly to acting with nature. I feel again that such intersections help us both to get poiesis into focus and to foreground what is strange and unfamiliar about it. But I should emphasize again that these nonwestern worldviews do not come first in this book. I do not start from Daoism or animism and build a picture around that. You do not have to be an animist to follow the argument. My accounts center on maneuvers in fields of agency, and the point is that different modern and nonmodern philosophies resonate with them in different ways.

9. *Hybridity*. A technical point which complicates the story without changing the plot. I have written so far about acting-with and acting-on as if they can always be cleanly separated and distinguished, and the first point to make is that this split works here. I have chosen examples that bring out clearly the key features of poiesis, my main concern in the book — I want to get this unfamiliar pattern of acting into focus. And to bring out the contrast, I likewise emphasize clear examples of enframing. But there is more to be said.

Consider enframing as a stance. We typically do attempt to impose our plans on the world. We design dams, say, and go out and build them to that design. However, that in turn inevitably throws up problems in practice not anticipated in advance — how to cope with the unanticipated peculiarities of this particular construction site, and so on. And these peculiarities can often only be handled in an ad hoc, poetic fashion — finding out what works here and now. In this way enframing and poiesis are chained together, like yin and yang. But the point I need to stress is that this does not imply a symmetric relation between the two. The plan continues to structure the overall project, with poiesis filling the gaps. We can say that in general, enframing has a *fractal* structure in which poetic adjustments are parasitic on the overall trajectory of acting-on. Much the same can be said about enframing technologies. They often do succeed in dominating their worlds, but they degrade in use and require maintenance and servicing, which can again take a poetic form, but the acting-with here is also parasitic on acting-on. Once more, then, it seems appropriate to describe enframing as a practical gestalt that foregrounds mastery against a background of poiesis and findings-out.

Coming from the other direction, I can note that in my examples, poiesis is entwined with enframing. For instance, as acting-with, the adaptive management of the Colorado River (chapter 4), depends on the enframing structure of the Glen Canyon Dam to act on and modulate water flows and experimental floods. In fact, I find it impossible to imagine any poetic project or technique that does not also somehow entail fixed and reliable elements of mastery like that. But what interests me about projects like adaptive management is precisely that they foreground nonenframing aspects, the aspect of finding out what the river ecosystem will do and acting with that. This gets us back to the metaphor of practical gestalts: despite their hybridity, poetic practices and techniques lean on becoming and emergence, just as enframing leans on mastery.[15]

I have discussed acting-with and acting-on in terms of different ways of standing in the dance of agency, and we could think of my examples as limits at the ends of a spectrum. I think this is a good way to get poiesis, above all, into focus. It is no doubt the case that we could find intermediate examples in which the entwining of acting-with and acting-on is more evenly balanced and speaking of gestalts would find less purchase. But from the perspective of this book, such examples would serve only to muddy the waters, which is why I do not explore them further here.

10. *Scope.* It is easy to fall into a totalizing idiom and to write as if the pairing of poiesis and enframing can exhaust all possible patterns of action, but that is not my intention. Certainly, at an individual level many of our actions are oriented neither to domination nor finding out and experimentation. I should therefore simply note that my concern in this book is with organized and repeatable engagements with nature (flood defenses, farming, and so on) that, more or less explicitly, revolve around themes like controlling (though not necessarily dominating), managing, and generally getting along with the nonhuman world.

. . . .

This book is not written as an academic argument. One familiar tack would be to elaborate, say, the ontological discussion of the dance of agency, relate it positively or negatively to the thought of other philosophers and social scientists, alive and dead, and so on — and thus to remain within the orbit of scholarly thought. I want to do something different. I want to conjure up and get clear on poiesis as an important but unfamiliar stance in the world, and the only way I know to do that is through examples — examples that tie the preceding remarks into the world — that show us what poiesis can look like in practice. The sequence of topics is as follows:

> CHAPTER 1: Eels. This chapter tells a story of an invasive species, exemplifying a view of the world as centered on action and performance rather than words and ideas, and thus offering a model for thinking about the following chapters. Themes include human and nonhuman agency, the dance of agency, and the domination of nature. The story's generality is suggested by a comparison with the war on terror and responses to the coronavirus.

> CHAPTER 2: The Mississippi. This chapter on flooding and river management exemplifies the same vision as chapter 1 but at a different scale. It discusses enframing as a particular stance in the flow of becoming, the relation between science and enframing, and connections between enframing and disaster. It touches for the first time on poiesis, which appears here as letting go.

> CHAPTER 3: Erosion Control. Our first example of poiesis in action. This short chapter sketches out a simple model of poiesis to be elaborated in succeeding chapters. It emphasizes the efficacy of poiesis, and also the absence of science in this case.

> CHAPTER 4: The Colorado. The focus is on the adapative management of the Colorado's ecosystem. Poiesis is examined here at greater length as a *process* of experimentation and adaptation, and as a *technique* in which the dance of agency becomes a choreography of agency. The chapter explores the contrast between poiesis and enframing, and the sense in which science is a fallible shortcut around poiesis. Two senses of "experiment" are distinguished — in the laboratory (science) and in the wild (poiesis). The chapter includes a discussion of scientific modeling, and of ways in which cybernetics illuminates key features of poiesis.

> CHAPTER 5: Water. This chapter sketches out the long-term evolution of water management in the Netherlands, and a paradigm shift away from enframing and toward poetic approaches, with the Room for the River

14

project and rewilding as examples. Christian and animist ontologies are contrasted in relation to enframing and poiesis, respectively.

CHAPTER 6: Natural Farming. The focus here is on a distinctive approach to "natural farming" contrasted with conventional approaches. Poiesis is foregrounded again as both process and technique, leading to another choreography of agency. The critique of science from natural farming is explored, as is the relevance here of cybernetics and Daoism.

CHAPTER 7: A Choreography of Fire. This chapter discusses the Aboriginal choreography of fire in Australia as poiesis and creative management, contrasted with conventional fire-control techniques and scientific burning. It also notes the relation of Indigenous knowledge and animism to poiesis in this instance.

CHAPTER 8: Spirits. Amazonian shamanism is analyzed as a poetic technique, rejected by science. This chapter significantly broadens the frame of the analysis by engaging with non-standard forms of agency entangled with technologies of the self. A novel form of dualism — symmetric, not asymmetric — appears. This chapter engages with the ontological turn in anthropology and science and technology studies, and with the possibility of different scientific and non-scientific worlds.

CONCLUSION: Poiesis. This summarizes the book's argument and key concepts and clarifies some important points. The book ends with further discussion of the Anthropocene, the politics of poiesis, and the question of what education for poiesis rather than enframing might look like.

1 Eels

The Dance of Agency

Who danceth not, knows not what is being done. —ACTS OF JOHN, Gnostic Gospel

This chapter offers a preliminary exemplification of some of the key concepts that run through the book: performance, agency, the dance of agency, decentering, and emergence. To anchor them in the world, we can consider a short story about an invasive species, Asian eels (Pickering 2005a).[1]

Starting more than twenty years ago, these eels were imported into the United States as pets, attractive additions to the usual tropical fish found in aquariums. When they arrived, the eels were quite small and cute, but they turned out to grow rapidly to several feet in length. Worse, they climbed out of their tanks. I imagine a scene like something from the science-fiction movie *Alien*. The little girl comes downstairs in her nightie to be confronted by a row of teeth leering at her over the edge of the fish tank. Horrified owners responded to this by dumping their eels in local ponds. But then it turned out that the eels could compete quite successfully for food with the local fish, especially in the southern United States, much to the annoyance of sport fishermen, who found that their catches of large-mouthed bass were decreasing. This annoyance was passed on to local authorities, who passed it on to civil engineers, and one response was to try draining the affected ponds. That did not work. The fish died before the eels, which, cleverer than the fishes, burrowed into the mud and waited for the ponds to refill. Another response was to try to confine the eels — to stop them invading major waterways — by building concrete barriers around the ponds. This failed too: the eels just climbed over the barriers.

That is the story (a glance at the web shows that the eels continue to spread). Now I want to explore it from several angles, picking out features that are foundational to what follows.

1   It is not a story about anything cognitive, mental, linguistic, epistemic. No doubt the people involved — aquarium owners, fishermen, engineers — did do some thinking along the way, but this does not to go to the heart of the matter. The center of gravity of the story is securely located in the nonlinguistic realm of action, performance, doings — the doings of eels and of people.

2   People and eels are very different biological beings, but at the level of performance the story is symmetric: the humans and nonhumans (the eels) are pretty much on a par as agents. The performances of both are integral to the story and neither is in control. We could say that this symmetry puts us in our place — in the thick of things, just a part of nature, not its masters.

3   This symmetry is more than a mirroring; it is also a constitutive coupling. The doings of the humans are specific responses to the doings of the nonhumans and vice versa. We bring the eels across the Pacific; they grow and scare the humans; the humans throw them out; the eels thrive at the expense of the fish; the engineers try to tilt the balance, draining ponds and building barriers; and so on and so on. This is an example of the kind of back-and-forth in performance that I call the dance of agency — the agency, in a performative sense, here of the humans and the eels.

4   The dance of agency is a key concept that runs through this book, and the eels story is intended to bring it to life and help us remember what it means. My thought is that it does not refer just to eels or just to invasive species but to everything. From a performative perspective, the world is made and remade not from quarks or black holes or whatever but in dances of agency. The eels, then, are a small mnemonic, a memory-aid, for this big overall ontological picture.

5   I have not properly argued for the claim that one can find dances of agency everywhere. But I first stumbled on this concept in the early 1990s in my studies of the history of physics, where I found myself saying that scientists engage in dances of agency with their materials, instruments, and machines (Pickering 1995a). And since then I have documented and analyzed dances of agency in all sorts of areas, running from mathematics and industrial production to the nineteenth-century rise of the synthetic dye industry and the intersections of science, technology, and warfare in World War II — as well as the many examples I wrote about in the history of cybernetics (Pickering 1995b, 2005b, 2010). In the end, it becomes hard to imagine how one would not find oneself

talking about dances of agency if one thinks of the world as a locus of consequential actions, which it surely is.

Two further observations on these dances are worth making. First, they are decentered; neither the humans nor the eels called the shots in their interaction or where they were going: each responded to the other. In contrast to an asymmetric dualism, this is a story of humanity as plunged into and carried along by a world we did not make and do not control. Dances of agency sweep like tornadoes across the chessboard of stable life.

Second, we can note that a specific sense of emergence is implicit in this story. Emergence here refers to the appearance of brute novelty in the world. No one could have predicted in advance how events would evolve from pet shops to eels burrowing into the mud surrounded by angry fishermen. Science, we could say, veils this sort of emergence from us by imagining a lawlike world in which everything is somehow predictable.

. . . .

The eels have done their work. They have provided us with a simple example of my understandings of agency and the dance of agency that we can hang onto and refer back to. In closing, I want to repeat my conviction that dances of agency are everywhere. There are plenty of examples in the following pages, starting with the Mississippi River in the next chapter. But to go to an extreme, I want to say that world history in general has just this quality. All that has followed from the events of September 11, 2001, the so-called war on terror, for example, has evolved as an unpredictable back-and-forth of actions and reactions—the collapse of the World Trade Center; the invasions of Afghanistan and Iraq; the rise and fall, but not disappearance, of ISIS; the conflagration of Kurds, Turks, Syrians, and Russians in Syria; proxy duels between Iran and the United States; and the continuing instability of Iraq and Afghanistan (which we have just left to the Taliban after twenty years, as of August 2021). No one controls this historical trajectory; no one can predict where it will carry us next. The trajectory of the coronavirus likewise fits the blueprint: the virus crossing over from animals to humans; its varied and sometimes lethal consequences for those infected; our various countermeasures including quarantines and the development of vaccines; reciprocal transformations of the inner being of the virus itself in the appearance of "variants"; and so on—again, a performative, decentered, and emergent story, in which we are by no means in control and with a destination that remains unknowable.[2]

But in the flux, we often long for control, and we often act as if we can achieve it. The humans did not engage in a dance of agency with the eels for the fun of it. They wanted to bring it to an end on their own terms. Draining ponds and building barriers have been parts of the search for a final solution in which we humans gain dualist mastery of the eels, either pinned down and fixed in place or simply exterminated. From this perspective, our interactions with the eels can be seen to display the stance of enframing and acting-on. We can explore that further in the next chapter, where there is more substance to hang onto, and where we can also begin to think about acting-with.

2 The Mississippi

Enframing and Letting Go

The previous chapter focused on the dance of agency and touched on enframing only briefly. This chapter reviews a weightier example of the dance and examines enframing at greater length, including its dark side. We can touch briefly on poiesis, too, but that is the main concern for the chapters that follow. The topic of this chapter is the management of the Mississippi River (Pickering 2008), and my original inspiration comes from John McPhee's book *The Control of Nature* (1989).[1]

The Mississippi is one of the world's great rivers. All the rain that falls in the Midwest of the United States drains through it into the Gulf of Mexico. Prior to European settlement, the lower reaches of the Mississippi were marked by natural embankments of sediment about three feet high — levees — deposited on either side of the waterway. The levees usually served to contain the river, though sometimes it would overflow them and inundate an enormous floodplain. It appears that the human inhabitants could live with that — the Native Americans would simply move to higher ground when the river overflowed and move back later.[2] We could speak of a simple and even graceful dance of agency between the river and the human population. The aim was not an enframing imposition of the human will on the river but rather a gearing together of human and nonhuman performances. Nomadism was thus a poetic stance of acting with, rather than on, nature — our first instance of poiesis in action. And we should not, in fact, think of this nomadic relation to water as entirely buried in the past. We will encounter a present-day version in the Netherlands in chapter 5 and, actually, a similar pattern can be found today around the Exe River where I live in southwest England. Usually cattle graze here on the rich grass of the river's floodplain — except when the river floods, as it often does, and the cows just move onto higher ground (Pickering 2013a).

But, returning to our story, then came the European settlers, who began to establish fixed towns along the Mississippi — notably New Orleans, founded in 1718 as the river's major seaport. Cities cannot come and go like nomads, so with the growth of these towns, the containment of the river became a matter of increasing importance, giving rise to one of the world's great projects of the domination of nature — nothing less than the control of the Mississippi itself. Integral to this was an artificial raising of the levees to confine the river within its banks. This made obvious sense as an engineering strategy; but what interests me here is that it periodically failed. As the levees rose, the river rose as well; flooding continued; the levees had to be raised further; and so on, back and forth, right up to the present. As a result, New Orleans became a walled city, surrounded by a ring of earthworks thirty feet high. McPhee compares it to the walled cities of the Middle Ages, though the enemy now is water, not the humans beyond the walls. Relative to the streets of New Orleans, massive cargo boats on the river now pass overhead. As McPhee says, if the levees weren't in the way, the water traffic would present a surreal spectacle reminiscent of an elevated railway.

So, the same general picture emerges here as in the eels' story, though now on a much bigger and more consequential scale. Again it is a story that revolves around actions and performances, the doings of the engineers and the river. And the form of the story is again that of a dance of agency, a performative back and forth of people and water. The engineers try something, building levees; the river responds, overflowing them; the engineers respond to that, raising the levees further; and so on, apparently ad infinitum. Neither the river nor the engineers are in control of this dance; it is a decentered process taking place in the interplay of the two. And the process is emergent as I have defined the term: its course is not given in advance, it maps itself out in real time. No one could have predicted the evolution of the river architecture and ships passing overhead.

We can continue the story in finer detail. For the past century and a half, responsibility for controlling the river has been assigned to the US Army Corps of Engineers, the ACE, which describes its work as a battle with the Mississippi — a battle in which the levees are central and of which the outcome is far from certain. It turns out, for example, that the Mississippi wants to move. It is now thirty feet above one of the lesser rivers it feeds into, the Atchafalaya. Left to itself, the entire Mississippi would sooner or later spill into the Atchafalaya, reaching the Gulf a couple of hundred miles west of its present destination and leaving the existing lower reaches of the Mississippi a mere trickle. This would be a catastrophe for cities like New Orleans that rely on the river water in all sorts of ways, and the ACE has accordingly been fighting the Atchafalaya for decades, reengineering its intersection with the Mississippi.

In 1963, a massive 250,000-ton weir became operational, designed to control the runoff from the Mississippi into the Atchafalaya and to prevent it exceeding its prior rate of around 30 percent. In the floods of 1972 and 1973, the control structure held, just. If it had failed, the Mississippi would have changed course irrevocably. After the flood, inspections revealed that the structure had suffered massive damage. Part of it had just gone: turbulent flows had excavated holes as big as football stadiums around it. Despite massive repairs, it would never meet its design specifications again. The original control project cost $86 million; after 1973, a new Auxiliary Structure was added, at a cost of $300 million, consisting of six gates, each sixty-two feet wide and together weighing 2,600 tons. McPhee quotes an engineer on the new project as saying at the time, "I hope it works" (1989, 52). Again we see the dance of agency playing out in the emergent and decentered intersection of the river and the engineers. This, I would say, is how things go in the world quite generally. Now I want to turn to the stance of the ACE, their way of conducting the dance.

. . . .

Man against nature. That's what life's all about. —GENERAL THOMAS SANDS, President of the Mississippi River Commission, quoted in John McPhee, *The Control of Nature*

The ACE does not relish the dance; the engineers do not engage in it willingly; it is a problem they want to bring to a close; they want to get out of it. The ACE's stated objective is to dominate and control the river, to bend it to its will. It describes its struggles with the river as a "battle" to fix the parameters of the river, "intending to control it in space and arrest it in time" and to "stop time in terms of the distribution of flows" between the Mississippi and the Atchafalaya (McPhee 1989, 7, 10, 21).

The ACE's guiding vision and objective is thus a future state in which the river is defeated; and all the problems along the way, the back and forth of the dance of agency, are backgrounded in that vision, veiled, forgotten. This, then, is a nice example of the *stance* of enframing — a specific structure or pattern of action which is superimposed on the dance of agency, aiming to subdue and domesticate the other and to make part of the world more dual, with us in charge.

And, as I remarked in the introduction, we can also understand the *technology* employed by the ACE to accomplish its ends as the ACE's proxy in enframing. The massive earthworks built by the ACE — the levees — are precisely intended *not* to respond to the river. However the river flows, the levees are intended to overpower the river's agency, to blot it out, and to prevent the water going beyond its prescribed course. Quite typically, as a technology of enframing, the levees are designed to be imposed on the landscape as a final solution to flooding.

I once thought this was obvious, how it has to be — "man against nature." How else could we act on a river other than dominate it and command it to serve our purposes (irrigating our fields, say, without flooding them)? The roots of this book lie, for me, in the realization that I was wrong. There are systematically other ways of engaging with nature (and machines, and other people) that are based on the alternative blueprint that I call *acting-with* and *poiesis*. The nomadism of the Native Americans mentioned at the start of this chapter already showed us a poetic alternative in living with rivers, and there are more examples to come.

. . . .

When uncertainties are wished away and not planned for, the crises that follow are all the more intense. —C. S. HOLLING, *Adaptive Environmental Assessment and Management*

The more efficient the technology, the more catastrophic the destruction when it collapses. —WOLFGANG SCHIVELBUSCH, *The Railway Journey*

The planetary crisis will manifest itself with exceptional force in those parts of the Earth that have been most intensively terraformed to resemble European models. —AMITAV GHOSH, *The Nutmeg's Curse*

The sum of evil, Pascal remarked, would be much diminished if men could only learn to sit quietly in their rooms. —ALDOUS HUXLEY, *The Doors of Perception*

In the introduction, I mentioned the idea that enframing has a dark side and that this is an important motivation for an interest in poiesis. Two lines of thought intersect here. One is that enframing cuts us off from the world in all sorts of ways (mentally, physically, emotionally, spiritually) and breaks any lively connection. This is clearly the case with the ACE's attempts to manage the river; to the ACE, the river is just a problem to be mastered. Of course, the river could be a problem to the Native Americans, too, but their poetic response was to be familiar with its ways and build its actions into their lives, not blot them out. This contrast between cutting off and intimate performative engagement is an important aspect of the difference between acting on and acting with the world — a reminder that whether we like it or not we are caught up in a lively world we cannot control.

From another angle, the suggestion is that while enframing can certainly bring significant benefits, it also brings increased chances of catastrophe and disaster. Wolfgang Schivelbusch (1986, 129–33), for example, noted that nineteenth-century railway accidents were much more violent than earlier technological disasters, a fact that he understood in terms of the exceptional "falling height" (153) of the energy bound up in a speeding train.[3] Something similar seems to have been happening with the Mississippi. The devastation of New Orleans by Hurricane Katrina in 2005 would be a case in point.[4] More generally, a recent study concluded that "the interaction of human alterations to the Mississippi River system with dynamical modes of climate variability has elevated the current flood hazard to levels that are unprecedented within the past five centuries" (Munoz et al. 2018, 95). In other words, even if one takes climate change into account, it seems that levee construction and other engineering works on the Mississippi have made the perils of flooding worse — and the record-breaking floods of 2019 point in the same direction (Renfro 2019). Here, then, the dance of agency takes the shape of an *arms race*, as the river continually raises the stakes for the engineers and vice versa.

This sort of arms race is evidently a self-generating lock-in from which it is hard to escape, just like a nuclear arms race or the war on terror, or drugs, or whatever. In a recent lecture, "In Praise of Floods," James Scott (2020) discusses the enframing (though he does not use the word) of rivers as *iatrogenic*, a continual attempt to remedy the consequences of prior interventions. We will see similar phenomena in later chapters. Conversely, of course, Native American nomadism left the world as it was, without precipitating any arms races — a less perilous way to be and another important contrast with enframing.

....

We have to unleash the power of the river, the river that built the Delta in the first place. —DAVID MUTH, quoted in John Schwartz, "A Mini-Mississippi River May Help Save Louisiana's Vanishing Coast"

This is an appropriate point to wonder briefly what poiesis as an alternative and antidote to enframing in river engineering might look like, before a more systematic examination in the following chapters. I have mentioned nomadism as a poetic engagement already, but how else could we do things differently?

I first wrote about the Mississippi around the year 2000, and I found myself wondering then about the *So what?* question. What difference might my analysis make for New Orleans, say, *in practice?* I had been taking it so much for granted that we have to defend it from floods and from the loss of the Mississippi to the Atchafalaya, that I burst out laughing in shock when it suddenly occurred to me that there was a very obvious alternative to enframing the river: to stop fighting it and let New Orleans go (Pickering 2000, 2001a). At the time no one else seems to have thought of that.[5] It appeared very harsh, of course, so I softened the blow and argued that people should have the option to move away from New Orleans when the weather was quiet, rather than waiting to be hit by a disaster that was bound to arrive sooner or later. After Katrina, I felt that I had been not harsh but prescient, and that "let New Orleans go" could be the slogan for a nonenframing stance in the world. This is, in fact, an example of the first poetic stance I mentioned in the introduction as the opposite pole from enframing. Instead of trying to dominate and indeed extinguish the agency of the river, we could surrender to it and, so to speak, let its agency shine through and structure the situation. We could go with the flow, to adopt an appropriate Daoist motto.[6]

I could add that while I was almost alone when I first thought of letting New Orleans go, very recently it seems that this strategy is becoming recognized under the heading of "managed retreat" (Mach and Siders, 2021a, 2021b). In 2022, the US Department of the Interior made grants of a few million dollars to five Native American Indian tribes to help them begin to relocate to higher ground, away from oceanside settlements threatened by coastal surges, flooding, and erosion (Flavelle 2022a, 2022b, 2022c).[7]

This sort of relocation in the face of environmental threats echoes the original Native American nomadism around the Mississippi as well as the farmers in the Exe Valley, and it will appear again in the Room for the River project in chapter 5. But this version of poiesis as surrender to the agency of the environment is too passive to be general. So it is interesting to think here about a recent more constructive real-world initiative in engineering the Mississippi. This concerns plans to build a "diversion" in the bank of the river, a set of gates in a levee in Louisiana that can be opened to permit out-flows from the river, in the hope that river sediment carried through it will rebuild the coastline that has been eroding behind the levees (Kolbert 2019).[8] Here, there is an active role for human agency (constructing the diversion and modulating the flows) and for the river, too (carrying the sediment and rebuilding the land). This explicit coupling of agencies, in contrast to the ex-tinction of one by the other in enframing, is a "complete about face" in river management (Brad Barth, quoted in Schwartz 2020, 3) and inclines me to call this another instance of poiesis.[9] It is interesting that the gates will only be opened on occasions when freshwater flows will not affect fish and oys-ter stocks too much (Schwartz 2020, 6).[10] This poetic coupling of human and nonhuman agencies in terms of timing and a balance of objectives is some-thing we will examine in more detail in another ecological restoration proj-ect, on the Colorado River, in chapter 4.[11]

．．．．．．．．．．．．．

One last topic before we leave the Mississippi. As mentioned in the intro-duction, I am interested in the role of science in our dealings with the world. This is discussed in more detail later, but here I just want to note a simple contrast. On the one hand, the ACE has not acted blindly in its struggles with the Mississippi. Plenty of science has fed into this instance of enframing. Sci-ence appears, in fact, in two guises in this story. Until the great flood of 1927, the ACE was guided by the best scientific hydrology available in acting out its policy of "levees only" (McPhee 1989, 41). The idea was that confining the river would force it to flow faster, cutting into and deepening its bed, and thus further decreasing the risk of flooding. Evidently, this did not work. As the levees rose, so did the river, continuing the dance of agency we have dis-cussed. And science was itself mangled, as I would say, in this process. After 1929, scientific modeling took over. Initially, this involved the construction of physical models of the Mississippi, the grandest of which — and surely the biggest analog computer ever built — was the Mississippi Basin Model, con-struction of which was begun in 1943 and completed twenty-three years later, covering an area of two hundred acres near Clinton, Mississippi (Cheramie 2011; Fatherree 2004). Engineers could control waterflows into the entire model or parts of it and measure responses — levels and flows — at different points downstream, helping them, for instance, identify places where flooding would be likely and where levees needed to be reinforced. The model, in general, helped make the river a "manageable object" (Cheramie 2011, 7). In the 1970s, use of the Mississippi Basin Model was phased out in favor of computer simulations of the river which serve the same purpose.[12]

In contrast to the scientificity of enframing, I can emphasize that little or no science or theoretical knowledge entered into the poetic relations to the river mentioned above. No science entered into the presettlement nomadism of the Native Americans around the Mississippi any more than it does in the performance of farmers and cows in the Exe Valley today. Nomadic practices hinge instead on a direct performative coupling between the inhabitants and the behavior of the rivers in question. The same can be said of the poetic strategy of letting go, as in my hypothetical example of New Orleans and in present-day managed retreat.

This aspect of the contrast between enframing and poiesis is important because of our tendency to think of science as the cure for all ills. There are other ways of going on which do not involve science at all, and this book is an attempt to get them into focus.

Having said that, we can muddy the waters a bit. Constructing diversions in the levees in Louisiana has itself been involved with a certain amount of analog and computer modeling as a guide to potential repercussions.[13] We can think more about the role of this sort of science in chapter 4.

....

This completes our first pass through enframing — another example to hang onto as we proceed. We have examined enframing from two angles: (1) as the dominating stance of the ACE in its dance of agency with the Mississippi River, seeking to negate the agency of the river and to bring the dance to a dualistic end, and (2) as the mode of functioning of technologies like levees, which serve to blot out the river's agency. We have also seen the dangers of enframing — both divorcing us from nature and precipitating an increasingly perilous arms race with the river.

From the other side, I have mentioned some instances of poiesis that cluster around the Mississippi — nomadism, letting New Orleans go with managed retreat, cutting diversions in the levees — and I have contrasted the presence and absence of science in these projects of enframing and poiesis. But these examples were tangential to my central concern, in this chapter, with enframing, and the following chapters aim to elaborate a much more detailed appreciation of poiesis. The next chapter sets out a very short but incisive example that we can hang onto as a mnemonic in subsequent chapters.

3 Erosion

Poiesis

The previous chapter focused on enframing as a specific stance in the dance of agency. I touched briefly on some examples of poiesis as an alternative stance, but this chapter begins our systematic examination of poiesis. Our example here is taken from James Scott's book *Seeing like a State* (1998, 327), where Scott outlines a traditional Japanese approach to forest management:

> Erosion control in Japan is like a game of chess. The forest engineer, after studying his eroding valley, makes his first move, locating and building one or more check dams. He waits to see what nature's response is. This determines the forest engineer's next move, which may be another dam or two, an increase in the former dam, or the construction of side retaining walls. Another pause for observation, the next move is made, and so on, until erosion is checkmated. The operations of natural forces, such as sedimentation and revegetation, are guided and used to the best advantage to keep down costs and to obtain practical results.

As usual, like the story of the invading eels and that of the US Army Corps of Engineers versus the Mississippi, this is a story centered on action and performance, human and nonhuman, and again it is a story of a dance of agency, as nice an example as one could wish for. The engineer does something — building some little dams, say; the water does something in response — flowing in rather different channels; the engineer's actions respond to that; the water responds in turn, in a performative back-and-forth, until erosion comes under control. Like the eels and the Mississippi, this story can serve as an emblem or mnemonic for a general performative ontology.

Nevertheless, Scott's story has a very different feel from that of the ACE and the Mississippi. The forest engineer and the ACE approach water very differently, and that is what we need to think about. The difference resides in the way the dance of agency appears in their practice. The ACE pays no explicit attention to the dance. They calculate how big and strong the levees need to be to contain, say, a hundred-year flood and build the levees accordingly (this is, of course, an idealization). No awareness of a dance figures here. As we saw, the dance only comes into existence once the levees have failed to do their job, when the cycle repeats — the ACE reluctantly goes to work again on the levees, and so on. This veiling or marginalization of the dance of agency is, I take it, characteristic of enframing.

Instead, in Scott's story, the dance is the explicit focus of engineering work. There is very little else to this approach other than deliberately staging the dance: the whole point of building check dams is to find out how the water will respond. And this pattern of deliberately staging dances of agency is what I take to characterize poiesis as a stance in the world (poetic technology requires some further thought in succeeding chapters).

This is the key contrast I wanted to make, and we can conceptualize it as a gestalt switch. Gestalts are often thought of in terms of the organization of visual patterns of foreground and background, associated with amusing gestalt diagrams in which one can see ducks or rabbits but not both at the same time, or an old woman or a young woman (Hanson 1958, chapter 1). The two ways of dealing with water here differ precisely in one of these gestalt reversals, except that foregrounds and backgrounds are picked out in different practices and performances rather than visual interpretation. For the ACE and enframing, the dance is in the background of awareness, only to be unwillingly noticed and reextinguished when the river floods again. In poiesis and this Japanese approach to soil erosion, the dance is instead at the very focus of attention, deliberately called into existence and explored. And science comes into this contrast in an obvious way. In the previous chapter, scientific principles like levees-only and analog and digital models of waterways reinforced the invisibility of dances of agency by conjuring up a knowable and predictable river. This approach to erosion control instead entails very little knowledge at all and certainly no organized science. All that is left at center stage is the unveiled dance — acting with nature and doing without science.[1] That is another key feature to carry away from this chapter. But a couple of further points are worth making.

. . . .

First, we can think again about the dark side of enframing and note that Scott's book is principally intended as a wide-ranging critique of scientifically inspired enframing. The book aims, as its subtitle puts it, to explore how "certain schemes to improve the human condition have failed," and Scott tellingly recounts how a range of well-intentioned, scientifically designed megaprojects have historically led to famine in agriculture and unlivable cities as unintended consequences. Once more science, enframing, and disaster are bound up together in Scott's studies.

But second, like me, Scott is interested in other ways to be in the world, which he discusses in a concluding chapter (which includes the erosion story) as examples of what he calls "*metis*." *Metis* could be a candidate for identification as poiesis, but there is a possible confusion here which is worth clearing up. Scott paraphrases *metis* as "practical knowledge" and associates it with "practical skills . . . know-how . . . common sense, experience [and] a knack" (1998, 311), and none of these terms grasp what I mean by poiesis. Knowledge, skill, and so on belong to a human individual and characterize that person's capacities. Poiesis, on the other hand, is relational, a performative stance in the world entailing deliberately staging dances of agency, not a kind of knowledge like common sense or skill or a knack. It is true that one gains skill (practical knowledge, etc.) by acting poetically, but one can act poetically without any prior skill—like the dam operators in the following chapter, in fact. We can return to this distinction in chapter 7.[2] Now we can turn to our first detailed example of poiesis.

4 The Colorado

Chapter 3 sought to conjure up a simple and memorable example of acting with nature, poiesis in action. This chapter aims to extend our understanding with a richer study, of an ecological restoration project on the Colorado River. As in the previous chapter, I want to explore poiesis as a stance in the world, but I also want to go beyond that in thinking about poetic technology or technique.

Since the restoration project began with a dam, I will start with a quick discussion of dams and enframing as a reference point, a contrast with what follows. I need to make four points:

1   A hydropower dam on the Rhine River was one of Martin Heidegger's ([1954] 1977) examples of a technology of enframing. The point for us is that the dam just stands there, unchanging and indifferent to the waterway it controls, imposing its will on it, defining the river as a source of electricity — acting on, mastering, and enframing the river.

2   We can also think about enframing not as material technology but as a human stance. The process of conceiving and building a dam displays a stance of enframing inasmuch as dams are designed and planned out in advance and then imposed on the landscape. Some math and science plus some engineering lore make it possible to calculate in advance just where and how high and how strong the dam needs to be to contain so much water, generate so much electricity, and so on.[1] Again, we see here how science is complicit in enframing, a way of knowing the future and acting out our mastery.[2]

3   In my introduction, I discussed disaster as the dark side of enframing, and we can see how that works out in this case. Emergence never goes away, it always bursts out somewhere; enframing is at most an imperfect shortcut around it. Dams usually work but they are not absolutely re-

liable. Writing this book has made me aware that dams around the world collapse more or less disastrously rather frequently. I suppose I always knew this but I never thought about it. Before, dam failures and similar events were to me just background noise in world history; now I see that they form a pattern as the dark side of enframing. I have been through a gestalt switch.[3] As I mentioned in chapter 2, Wolfgang Schivelbusch (1986) conceptualized the increasing perils of enframing in terms of the "falling height" of the systems in question, and dams illustrate his point: the bigger the dam, the bigger the disaster when it fails.

4   From another angle, we could also think about what I call the performative excess of technologies of enframing — the idea that besides accomplishing their intended purpose, they often do something else as well (Pickering 2017b).[4] Besides controlling the waterflow, the Glen Canyon Dam that figures below led to unintended and unwanted transformations of downstream ecology — a different sort of dark side from brute failure.[5] The restoration project discussed here was an attempt to address these transformations.

....

Completed in 1963, the Glen Canyon Dam on the Colorado River is situated upstream from the Grand Canyon. Like all such constructions, it set in motion unintended and unwanted downstream physical and biological transformations. Sandbanks eroded; native species of plants and fish became endangered; alien species started to move in. Here, we can examine attempts to do something about that, to restore the downstream ecosystem to something like its previous condition. Unlike dam construction, there was no way of calculating how to do that — the stance of enframing, ordering nature about, was not an option. We will be concerned instead with a project of so-called adaptive management as our example of poiesis, a way here of feeling our way through a "wicked problem," as mentioned in the introduction.[6] I take my lead from Lisa Asplen's pioneering study (2008) and more recently from James Rice (2013).[7]

The starting point for the Glen Canyon Dam Adaptive Management Program, the AMP for short, was an artificial flood that happened in 1983. In that year, Lake Powell, upstream of the dam, was in danger of overflowing, and as an emergency measure vast quantities of water were released through the dam from the lake into the river. And it was noticed that that this had beneficial downstream effects: "Depleted beaches were replenished, exotic vegetation was killed along the riverbanks, and previously degraded animal habitats were re-created" (Asplen 2008, 172). Taking inspiration from this, at the heart of the AMP that followed was a succession of deliberate artificial floods, known as High Flow Experiments (HFEs), modeled on this first one, each aiming to learn from and adapt to (hence "adaptive management") the upshot of the others in terms of operating conditions and downstream modifications.[8]

To date (January 2024), nine HFEs have been conducted on the Colorado River — in 1996, 2004, 2008, 2012, 2013, 2014, 2016, 2018, and 2023 (GCDAMP Wiki, 2023) — and in this case, unlike James Scott's example in chapter 3, nature has not been "checkmated." Early experiments looked promising. The first HFE, in 1996, initially appeared to be a success and "increasing sedimentation was readily apparent. . . . What we found was really quite extraordinary. The success of the event exceeds . . . even our most optimistic hopes of our staff of scientists" (Rice 2013, 418–19). But the optimism soon evaporated. Sandbanks that had been built up during floods eroded again; invasive species turned out to be hiding in the mud (like the eels in chapter 1) and reappeared; the population of humpback chub, an endangered species, continued to decline. "After two HFEs [1996 and 2004] and nearly 10 years of adaptive management, the ecological conditions below the Glen Canyon Dam exhibited little sustained improvement. . . . Long-term increase in sedimentation had not been achieved, the presence of humpback chub continued to decline even as nonnative trout were thriving, and nonnative vegetation, particularly tamarisk, remained entrenched along the Colorado River corridor" (420).

Nevertheless, "a great deal of insight into the interplay of controlled flooding and ecological response was gleaned after two HFE's," and it is interesting to reflect on that. The basic understanding of what was going on had to do with the sediment carried along by the river. The Glen Canyon Dam slowed the Colorado down, leading it to deposit most of its sediment above the dam, and the conviction arose that it was a lack of sediment that was at the heart of the problems below the dam — the river was washing away sandbars and beaches without supplying enough sediment to replenish them.[9] And the idea that came to the fore after the second HFE was that perhaps the sediment carried by two other rivers that joined the Colorado below the dam, the Paria and the Little Colorado, could be exploited. In particular, the idea was to wait until one of these rivers was in flood and thus carrying much more sediment than usual, and then to stage an experimental flood on the Colorado itself to carry this added sediment downstream, possibly recreating the sandbars and so on. The spring 2008 HFE was therefore "precipitated by flooding on the Paria River and an estimated sediment inflow to the Colorado 3 times greater than observed before the 2004 HFE. . . . There was evidence of redeposition of sediment. . . . Six months later, however, a majority of new sedimentation was lost to erosion, and [again] nonnative . . . vegetation was temporarily buried rather than scoured from the Grand Canyon. . . . Researchers now [2013] suspect that too much time had passed between sediment inflow and the 2008 HFE such that a substantial proportion was lost under normal operating conditions," and later HFEs were accordingly planned to be "triggered" by large sediment inflows (Rice 2013, 420, 422).[10]

This synchronization of flooding of the Colorado and especially the Paria became the pattern for subsequent HFEs and proved effective. "The evidence from [HFEs up to 2014] indicated that releases timed to follow sand inputs . . . are, in fact, an effective sandbar-building strategy. . . . Resource managers . . . consider the 2012–2014 results encouraging" (Grams et al. 2015, 2, 5). The sandbars continue to fade away between HFEs, and the current wisdom is that there need to be relatively frequent artificial floods timed to coincide with major sediment inputs from the Paria and Little Colorado. HFEs are now, in fact, routinely seen as helping "rebuild beaches, sandbars and other environmental resources" (United States Bureau of Reclamation 2018, 2).[11]

The basic point to note is that this story has just the same form as Scott's story of erosion control in the previous chapter, though on a much bigger, richer, and more consequential scale. In the foreground again is a deliberately staged dance of human and nonhuman agency. The dam operators repeatedly release floods of water through the Glen Canyon Dam to find out how the river and downstream ecosystem will act in response and tune their subsequent actions in response to that, and so on, back and forth. The AMP is thus our first extended example of poiesis now in contemporary river management and ecological restoration.[12]

. . . .

So far, we have discussed the AMP as an example of the poetic stance in the world, a disposition to act in a certain way, staging dances of agency to find out where they will lead. Now we can turn to a topic we have not considered before (apart from the brief mention of nomadism, letting go, and diversion on the Mississippi in chapter 2) and which will reappear in the following chapters: the AMP as an established poetic technology or *technique*.

In his 2013 account of the AMP, James Rice (2013) described the program as a "failure," a "failure to achieve sustained resedimentation and thus ecological alterations" (423), and this sense of failure is clearly relative to the ambition of enframing the river. We could imagine that a specific experimental flood might have permanently reconfigured the downstream ecosystem as desired. That would be a classic final solution; nature would have been checkmated, as Scott said (misleadingly, I think) of erosion control in chapter 3. In fact, as we saw, gains have tended to fade away — invasive vegetation has re-emerged over time, and so on, and this is the sense in which the AMP and HFES have failed.

But we can look at this from a different angle. The AMP can be seen as moving at least towards not a one-off solution but a repeatable procedure — a technique more than a technology — for managing the river and its ecosystem: synchronized flooding as a continuing process of ecological restoration, in which artificial floods are repeatedly released from the Glen Canyon Dam to rebuild sandbanks just when floods on the Paria or the Little Colorado Rivers are carrying extra sediment into the lower reaches of the Colorado itself. And what interests me most here is that this technique is itself a very nice example of *acting with*, rather than acting on, nature. The technique does not impose itself on nature come what may like a dam but takes advantage of the found liveliness of nature, gearing the human agency of the dam operators into the nonhuman agency of three rivers (Colorado, Little Colorado, and Paria). We could think of this gearing together as a *choreography of agency* — the establishment of a regularized score, say, for a dance of agency in which nature both leads and follows.[13] And this is the sense in which synchronized flooding is a poetic technique. It is, so to speak, part of or within nature rather than a technology that acts on it.

I am tempted to say that poetic techniques like this are an important discovery of this book. I already knew about poiesis as a stance in the world, but not as technology or technique. It seems to me there is something new and important here — a way of being in and relating to a lively world that modernity has never got sufficiently into focus. We are all very familiar with the way enframing technologies like dams and levees and seawalls dominate their environments, but I have never seen the choreography of agency discussed, certainly not in a scientific or engineering context. It is not clear to me that modern engineering or modern common sense has any image, even a word, for such a dynamic balance. The existence and efficacy of this way of engaging with nature is something it is hard to foreground, no doubt because it is a joint achievement of a combination of humans and nonhumans — not the sort of thing science (or the humanities, for that matter) equips us to think or speak about. The easiest way to get it into focus is to leap ahead to the end of chapter 6, where the Chinese concept of *shi*, the propensity of things, appears. Then I can say very simply that adaptive management is about exploring the *shi* of rivers, sand, vegetation, and so on and finding ways to combine them toward a desirable upshot.

And here it is worth reiterating that the choreography of agency is a performative achievement, a choreography of performances. It is not centered on knowledge or science. It is a sort of balancing of multiple agencies, in which the balance point (or zone) is located not by prior calculation but in the poetic exploration discussed in earlier sections — explorations which I also emphasized are centered on performance, not knowledge. The upshot then is that poiesis as both stance and, now, as technique, is "doing without science."

....

I want to emphasize the centrality of performance, worldly action, to poiesis, which is why I have stressed that our previous examples have entailed little or no science or articulated knowledge. Poiesis is just different in this way from our usual technoscientific approaches to enframing. The same lack and even critique of science remains the case in the chapters that follow. But in this sense, the adaptive management program on the Colorado is an exception: it has had considerable scientific input. I therefore want to think more about this, and about what sort of science is involved in the AMP. I can think of two angles from which to approach this question.

First, it is revealing to think about an ambiguity in the word "experiment." The AMP itself was certainly an experiment. It aimed at finding out about the behavior of the river and its ecosystem. But we should note that that is a different sense of "experiment" from the sense usually associated with science. Scientific experiment is typically associated with a *detour* away from nature in the raw and into the laboratory (or maybe to an outdoor scale model, as in chapter 2). There scientists can control the relevant variables in order to produce knowledge that can then be moved back to the control of nature. No such detour was involved in the AMP. There, the experimentation was directly on the thing itself, the Colorado River. We can think of this as an "experiment in the wild," and we should note that the principle concern of that sort of experimentation is with the performative outcome — how will the ecosystem behave? — not the production of transportable unsituated scientific knowledge.[14] This is a sense in which adaptive management lies in a different space from modern science.[15]

Having said that, however, we can think about the kind of science that has indeed figured in the AMP. I have long been accustomed to thinking of science (and technical knowledge in general) as something that comes first in practice — the established knowledge of volumes, pressures, and strengths that structures dam construction, for example. This is the familiar sort of science that sets the world up for enframing. And the point to emphasize now is that the science of the AMP is not like that. On the one hand, a considerable amount of monitoring and mapping has been involved. I have spoken so far as if the ecological effects of the Glen Canyon Dam and succeeding HFEs have been straightforward and obvious, but they are not. John Schmidt et al. (2001) report at length on the instrumentation and detailed measurements and calculations involved in mapping the many changes precipitated by the 1996 controlled flood along the four-hundred-kilometer reach of the river from the Glen Canyon Dam to Lake Mead, which I summarized earlier in this chapter in much less nuanced form. This sort of science makes the performance of the river and its ecosystem visible, which might be a preliminary to enframing but in this case feeds into the poetic AMP.[16]

On the other hand, much of the scientific work that figures in adaptive management entails the construction of mathematical models and computer simulations, "a kind of laboratory world" as one of its founders puts it (Holling 1978, 146). Thus, for example, "the LTEMP [Long Term Experimental and Management Plan] HFE Protocol uses predictive models to make recommendations for the magnitude and duration of potential HFEs using real-time measurements [another monitoring project] and models of sand inflow from the Paria River . . . to determine whether suitable sediment and hydrology conditions exist for a high-flow experimental release" (GCDAMP Wiki 2023).

What can we say about this? The point to emphasize is that though scientific modeling is in fact a conspicuous feature of the AMP, it plays more a modest, ancillary role as a sort of temporary conceptual scaffolding than the constitutive role of physics and engineering in projects like dam design. Thus the proponents of adaptive management both emphasize the utility of models and stress their limitations in environmental management. As Kai Lee (1991, 58, original emphasis) once put it, "*Surprise is unexceptional*. . . . Predictions are often wrong, expectations unfulfilled, and warnings hollow."[17] Given that, one might wonder what purpose modeling serves here — a question to which C. S. Holling and Shana Sundstrom (2015, 16) offer an answer: "Without modeling, practitioners feel they are operating in a barren world of inadequate knowledge and conflicting explanations."[18] We can see modeling, then, as an explicitly fallible technique for seeing in the dark, so to speak — better than nothing in terms of orientation but always revisable in practice (and see the discussion of Project Cybersyn in the next section).

Where does this leave us? It helps us see the forest for the trees. It reminds us that the poetic center of gravity of the AMP lies in the performative maneuvers in fields of agency we have been discussing, with scientific models arranged around them as tentative guides. To speak of "doing without science" would be an exaggeration in this particular case, but it points in the right direction if we want to understand how science can relate to poiesis.

. . . .

What we know of social, economic, and environmental behavior is much less than what we do not know. Therefore, the opportunity to benefit from chance and the unexpected should be part of the design goal. . . . The unexpected is to be expected. . . . We must ride with ecological forces as much as with social and economic ones. . . . It is possible to achieve designs that work with rather than against natural forces. —c. s. HOLLING, *Adaptive Environmental Assessment and Management*

I have so far emphasized that the center of gravity of the AMP is performative, and in that respect, science is an adjunct. But I noted in the introduction that there is an odd science (or field or approach) that hangs together with and illuminates not enframing but poiesis — cybernetics. I see cybernetics as part of a different, poetic paradigm from classical sciences like physics or chemistry (Pickering 2010), and I want now to think about how cybernetics relates to this discussion of the AMP (a theme we can develop further in chapter 6).

At the most fundamental level, the defining feature of the branch of cybernetics I discussed in *The Cybernetic Brain* was its ontology of unknowability — the formative idea that we will never fully know the world we live in, the fact that it can always surprise us; "surprise is unexceptional," "the unexpected is to be expected." It follows directly from this that we can never fully master and enframe the world, which is, of course, the realization that adaptive management puts into practice. If the world can always surprise us, the best we can do is try things out, see what happens and adapt to that — which is adaptive management in a nutshell.

In more detail, the cybernetic resonances of adaptive management seem clear enough. Though he does not figure in the history of the Colorado, the key figure in the overall development of adaptive approaches to environmental and ecosystem management from the 1970s onward was C. S. Holling, and in an early paper (Holling 1973; see also Holling 1978), he develops an analysis of ecological "resilience" in terms of phase-space diagrams, basins of attraction and tipping points that are recognizably cybernetic (Ashby 1960). Holling (1978) then presents a range of detailed studies of the adaptive management of several ecosystems, and Holling and Gary Meffe (1996) extend that picture to criticize "command and control" (enframing) in favor of adaptive environmental management as exemplified here. Adaptive management and cybernetics are, then, in the same space.[19] As in my earlier work on cybernetics, what my analysis adds to general discussions of adaptive management is precisely to pick out the *performative skeleton* of adaptive management from the details and scientific models of specific projects (Pickering 2021).

From a different angle, a rather technical observation might help in clarifying the overall contrast between enframing and poiesis. British cybernetics in particular grew out of a machine built by Ross Ashby in 1948, the homeostat (Pickering 2010, chapter 4). This was a device that interacted performatively with its environment, which was physically modeled by other homeostats, and reconfigured its own inner workings until equilibrium was reached. Ashby experimented with two different combinations of homeostats: a symmetric one in which all the homeostats adapted to one another and an asymmetric version in which the parameters of one homeostat were fixed and the other homeostats had to adapt to that. I find it helpful to see the asymmetric setup as a model of enframing. Thus, for example, waterflows and ecosystems had to adapt themselves to the Glen Canyon Dam as a fixed technological artifact. And, likewise, the symmetric setup can be understood as a general model of poiesis, in which each homeostat acts with rather than on its partners. In this sense the dam operators in the AMP function as a homeostat among other homeostats in collectively finding an ecologically productive balance point.[20]

From yet another angle, Stafford Beer's approach to management cybernetics made the role of calculational models in poiesis more explicit than the AMP, but still along the lines just laid out. In Beer's Viable System Model (VSM), organizations were understood in terms of five levels running from production through planning to higher management, and the key feature of the VSM was that all these levels together with the organization's environment were thought to be entangled in homeostat-like performative arrangements. Management, for instance, could propose organizational changes to the planners, but the planners could then evaluate these and make new proposals back to management, and so on, in a pattern that Beer called "reciprocal vetoing."

Interestingly, however, the performative skeleton of this back-and-forth was engaged with a set of a calculational models. In Project Cybersyn in the early 1970s — a cybernetic control system for the Chilean economy envisaged under the socialist Salvador Allende regime (Pickering 2010, chapter 6; Medina 2014) — a Systems Dynamics computer model of the world economy, for example, was imagined as informing management of the possible consequences of their decisions. Importantly, this model was not seen as running the show and dictating decisions, but it was rather to be continually revised in practice according to how well or badly its predictions matched observed behaviors. As I said of adaptive management in the previous section, then, this most famous of cybernetic projects explicitly saw calculational models as tentative and revisable adjuncts to poiesis — in the plane of poetic practice rather than controlling it from above. Beer caught the spirit of this in the title of a 1969 talk, "The Aborting Corporate Plan" — in which plans and models are framed as revisable and rejectable aids to seeing in the dark, subservient to action. The Viable System Model and Project Cybersyn are thus cybernetic mnemonics for the general role of models in poiesis.

Finally, I want to dwell on the connection between the present study and the body of work surveyed in *The Cybernetic Brain*. At one level, that book is all about poetic cybernetic techniques and technologies, in all sorts of fields, that adapt to their circumstances in practice, acting with rather than on their environments, just like synchronized flooding on the Colorado. In that sense, the present book extends my exploration of the poetic paradigm beyond my earlier work (and vice versa). On the other hand, it is worth saying that the examples in this book are "bigger" than those of *The Cybernetic Brain*, which featured things like adaptive artworks or little robots or a diagram of an adaptive organization. Here we are dealing with great rivers like the Mississippi and the Colorado. I hope that this extension makes it possible to take poiesis more seriously as real-world engineering. At the same time, the word "cybernetics" is itself a bit daunting, and I hope these stories of dams, levees, and so on are more familiar and accessible. And, of course, the subject matter itself is new — the environment is an important topic in its own right that did not figure in *The Cybernetic Brain*. Nevertheless, I see the studies in both books as filling out a picture of a single poetic paradigm. I think this is worth stressing.

....

It is interesting to contrast enframing and poiesis in terms of the dangers they pose. In my introduction, I mentioned two senses of danger we can associate with enframing. First, from a phenomenological angle, enframing acts to cut us off from nature, to decouple us from the world. Exemplifying this disconnection, the Colorado ecosystem was not even an active consideration in the construction of the Glen Canyon Dam. In contrast, we have seen that the AMP depended on a dense and attentive engagement with at least some aspects of the downstream environment, finding out how it would react to artificial floods of differing intensity and frequency, and thus bringing the liveliness of the river home to us. We can make another connection to cybernetics here. In 1958, Gordon Pask distinguished between what he called the scientific and the cybernetic methods. His argument was that the cybernetician aims to "maximise" interaction with an assemblage, while the scientist aims to "minimise" it (controlling variables in hypothesis testing) (Pask 1958, 171–73). Evidently, releasing megatons of water into a river, as in the AMP, corresponds to Pask's idea of "maximal" cybernetic interaction. And Pask thought of such interactions as a "conversation" — a nice metaphor for the open-ended back-and-forth of adaptive management.

Second, I have noted some of the physical dangers of dams as technologies of enframing — catastrophes and environmental degradation — and we have discussed adaptive management as addressing and ameliorating the latter. But what about the dangers of poiesis? Clearly, the poetic engagement of the AMP with the Colorado did not entail risks on the same scale as collapsing dams. But neither was the AMP entirely devoid of risk and we should think about that. Especially, it was recognized in advance that artificial floods might have an adverse effect on precious species, as had the dam itself.[21] So, are enframing and poiesis on a par as far as danger is concerned?

I would say not. Dam construction is a one-off, more or less irreversible, activity. You build a dam, find out what the effects are, and then you have to live with them. That is characteristic of enframing.[22] In contrast, poiesis has an importantly tentative character. Adaptive management is nothing but trial and error, trying this or that and reacting to what comes back. Unfortunate outcomes are indeed possible, but they show up incrementally and there is no need to repeat them.[23] And in the opposite direction, adaptive management can also seize on and accentuate positive outcomes — ameliorating ecological degradation in a way that an already finished dam cannot — which is the antithesis of endangering. My conclusion is thus not that poiesis is totally risk free, but that it is less risky and dangerous than enframing.

....

Two brief addenda to this chapter. First, it turns out that there is an enframing alternative (or complementary) approach to the AMP, designed though not implemented (yet), which it is illuminating to think about. This is to physically transport sediment from above the Glen Canyon Dam to below it. The simplest and cheapest version of this scheme is to use two massive dredges operating twenty-four hours a day all year round, to excavate three million tons of sediment from Navajo Canyon, thirty-three miles above the dam. The sediment would then be mixed with water to create a slurry, to be pumped in turn by three floating pumping plants and three land-based ones through first a submerged pipeline in Lake Powell and then up and over the Glen Canyon Dam, and into the Colorado River, where annual artificial floods would again distribute the sediment to downstream sandbars. The capital cost of this project was estimated at $220 million in 2006 dollars, with an annual operating cost of $6.6 million (Randle et al. 2007, 4, 8, 9, 12, 18, 25, 37). The point to note is that this scheme would dispense with the choreography of agency we examined above, replacing it with a human-centered technique imposed on nature — enframing once more, a final technological solution to the problem of the lack of sediment in the Colorado, operating ceaselessly on its environment in a relation of dualist domination.[24]

The contrast that strikes me here is this: the poetic approach to restoring the Colorado River ecosystem is imaginative and displays a certain grace and style. It is poetic in an everyday sense. The enframing approach, in comparison, would have to be described as "clunky." Scientists sometimes speak of "brute-force" methods where one grinds through endless obvious calculations for lack of any clever alternative, and I think that is how the sediment pipeline stands in contrast to synchronized flooding and how, in general, enframing stands in relation to poiesis.[25]

Second, I can note that to exemplify his foundational discussion of poiesis, Heidegger ([1954] 1977) pointed to the arts of ancient Greece. Like many people, I find this not an awful lot of use in thinking about the present — I cannot see how to bring this model into play in the world today. But writing this chapter has made me realize that, like Heidegger, I also find myself going back in time. The focus of the AMP is on precisely the restoration of the Colorado River ecosystem to a prior condition, and the mechanism for accomplishing this is to restore the periodic floods that in fact characterized the river before the Glen Canyon Dam was built. The AMP thus takes us back to the early 1960s and beyond.[26] In chapter 2, the diversion on the Mississippi was likewise a partial restoration of the status quo of river flows before levee construction, hoping to restore some of the Louisiana coastline. Scott's example of the erosion control engineer in chapter 3 was of a "traditional" practice, presumably out of fashion today. Rewilding (chapter 5) attempts to undo species extinctions; Fukuoka's natural farming (chapter 6) takes Japanese farming back thousands of years; Aboriginal fire (chapter 7) dates back a few centuries to the period prior to European settlement of Australia; shamanic communication with spirits (chapter 8) just has nothing to do with "modern" time zones.[27]

What is happening here? My conclusion is that in specific ways in specific practices we have collectively drifted into specific molds for enframing when we could have taken more poetic paths which have specific aspects to recommend them over their counterparts. I find myself, then, picking up poetic threads we have more or less dropped to see how they can figure in the present. I did not intend this in advance, though in retrospect, it is hard to see how it could have worked out otherwise.

. . . .

We have come to the end of our first sustained example of poiesis. We began with dam construction as an enframing contrast with my principal interest, the Adaptive Management Program on the Colorado River as a study of poiesis in action. I discussed the AMP as a poetic stance in the world, a way of engaging with the Colorado ecosystem, and then moved on to the establishment of a poetic technique, a choreography of agency: the pattern of synchronized flooding that emerged in and from the AMP. This technique exemplifies the notion of acting with the world, coupling human agency into the coordinated agency of several rivers in ecosystem restoration.

I sought to show how modern science fits in with enframing while remaining an adjunct at most to poiesis, which foregrounds performance instead. I tried to get at that contrast by thinking about two senses of "experiment" — in the laboratory (science as a detour away from nature) versus "in the wild" (adaptive management as performatively finding out how the river will behave). As well, I explored the role of calculational models and simulations in the AMP, as revisable guides wrapped around a performative skeleton. I then tried to show how cybernetic concepts serve to illuminate key features of this example. They help us grasp what is going on in it and, at the same time, to make transverse connections to a very wide variety of quite different cybernetic projects. They help sketch out a larger nonmodern paradigm of which our dealings with the environment can be a part.

At the end, I compared the dangers associated with enframing and poiesis; I mentioned the brute-force quality of an enframing alternative to the AMP; and I noted that writing this book has led me to an interest in rewinding history and following forgotten paths of acting with nature. The next chapter takes a broader look at changes in water management.

5 Water

A New Paradigm

....

Water figures prominently in the preceding chapters, so it seems appropriate to take a wider view now. The aim of this chapter is to show that the shifts in practice we have looked at are part of a more encompassing paradigm shift from enframing to poiesis. An exhaustive treatment is unnecessary and, anyway, beyond me. But for centuries the Dutch have been the acknowledged masters of water engineering — the model to which others aspire — and what follows is a brief review of shifts to poiesis in Dutch water practices. We can touch again on the connections between science and enframing, but as usual, the aim is to thicken up our grasp of poiesis. At the end, we can also look at some wider ontological and spiritual entanglements of enframing and poiesis — a topic we can take further in later chapters.

....

The Netherlands is a flat, low-lying country, much of it below sea level and prone to flooding. Put very simply, we can distinguish three epochs in which the Dutch have engaged with water in different ways.[1]

First, long ago, it seems that the inhabitants of the Netherlands lived in phase with nature, moving nomadically to patches of artificially higher ground known as *terps* when the lowlands flooded and moving back when the flooding receded, just like the Native Americans around the Mississippi and the cows in the Exe Valley. As before, we could speak of acting with nature and of a poetic choreography of agency, in which human performances were closely coupled into those of rivers and sea.[2]

Terps were built until the eleventh century, but from AD 800 onward, nomadism began to be displaced by a different engineering paradigm, the one the Dutch have long been famous for, drainage, and the familiar project of dominating and enframing nature was set in motion. Dikes (levees) and ditches were built and dug to drain the land and "reclaim" it, challenging it forth, in Martin Heidegger's terms, as standing reserve for settlement and farming (Zwart 2003, 112). Fixed farms and villages began to be established, creating in turn the need to defend them from flooding (like New Orleans and the Mississippi, chapter 2), and another kind of dance of agency was set in motion. Reclamation itself resulted in a lowering of the surface of the land at the same time as sea level rose, and "a series of dramatic floodings, known as the late medieval transgressions, created a sea (the Zuyderzee) in the very heart of the country" (Zwart 2003, 112). At this point, technological innovation entered the story. From 1408 onward, windmills were used to pump water out of reclaimed areas known as polders in conjunction with the building of canals and sluices to carry the water away.

Both senses of enframing are evident in this transition. Enframing as a human *stance* — the disposition to dominate nature and reorder it for human purposes — set in motion a dance of agency which in turn evoked a material *technology* of enframing (dikes, mills, pumps, canals) that accomplished the domination, separating land and water on behalf of its human creators. The last thousand years of Dutch history have thus been one long history of enframing, which has, of course, never been complete. Water has never been totally subjugated, so the dance of agency in the Netherlands, as on the Mississippi, has taken the form of an arms race, with nature and the engineers continually raising the stakes for each other.

We can come back to that, but first we can just note that science once more enters this story. Zwart (2003, 113) records that the mathematician Simon Stevin "worked on mills, sluices and canals and with the help of extensive calculations he was able to improve the windmill," including the use of a scoop wheel. The scoop was itself displaced by the Archimedes screw introduced by Symon Hulsebosch in 1634, and so on. It is interesting that here the detour through science and math does not refer directly to the behavior of water (compare the models and simulations of the Mississippi and the Colorado, chapters 2 and 4, respectively) but to machines like watermills and pumps. This marks a difference, I think, from Heidegger's idea that science sets nature up for enframing. In this instance, it is rather that the technologies of enframing themselves became a site for productive detours through science, another loop (chapter 4) away from and back to the things themselves.[3]

Battles with water continued from the Middle Ages all the way through the twentieth century. In 1953, a major flood caused the death of 1,835 people and 200,000 farm animals and set in motion very significant developments. The first reaction in the Netherlands was a continuation of the arms race with nature, an intensification of enframing and the construction of yet more hard defenses. The so-called Delta Works project was intended to shield the Netherlands better from the sea by closing off tidal outlets between Rotterdam and Antwerp so as "to reduce the chances of a future flood disaster to virtually zero" (Zwart 2003, 121; Bijker 2007 gives the intended odds as 1:10,000) — a classical dualist and dominating solution. But all did not go smoothly. Flooding continued despite the Delta Works, most notably in the floods of 1993 and 1995 when some 200,000 people were evacuated (McVeigh 2014, 1). At that point it became doubtful whether enframing and hard defenses could keep pace with the rising sea level associated with global warming. On the other hand, there was also increasing recognition of the ecological degradation that came along with hard defenses (Zwart 2003, 121) (paralleling ecological concerns about the Mississippi and Colorado as in chapters 2 and 4).

In the face of this conjunction, Dutch water management began to undergo a "paradigm shift" (Van Alphen 2020, 309; Vera 2009, 31; Lorimer and Driessen 2014, 173) from enframing back to poiesis. After centuries of work, money, effort, and lives devoted to dominating the rivers and seas, the Dutch began a qualified surrender, accommodating themselves to the agency of nature in all sorts of ways. To map out the new poetic paradigm we can look at three examples of what this transition looked like in practice. All these echo to some degree my idea of "letting New Orleans go" (chapter 2). Instead of trying to blot out the agency of nature with dikes and pumps, the Dutch have begun to go along with it, trusting it, letting nature call the shots.

1. "In 1997, a nick or groove . . . was made in a dune area near the town of Schoorl . . . in order to allow a more natural, more dynamic coastal landscape to develop, where wind and water reign. The sea enters the tidal inlet five to ten times a year, so that drifting, calcareous sands may change the properties of the soil, allowing unique forms of vegetation, typical for dynamic dune systems, to reappear and flourish. The ecological value of these border-line ecosystems is now emphatically recognized" (Zwart 2003, 123).

We can see this nick as a partial surrender to nature. For most of the year, the dune is a block, a classic technology of enframing, indifferently dictating terms to the water: thou shalt not enter. But when the tide is high enough, the water can act, flooding into the inlet and restoring, to some extent, its ecosystem (just like artificial floods on the Colorado and the construction of diversions on the Mississippi). An alternation between acting-on and acting-with, between enframing and poiesis.

2. Between 2006 and 2015 the massive (€2.2 billion) Room for the River project staged a similar maneuver but at a much larger scale. Covering four major rivers, the Rhine, Meuse, Waal, and Ijssel, the aim was to reduce the intensity of flooding by letting the rivers go — deconfining them, reducing the height of selected dikes, for example, and moving them farther away from the rivers, thus leaving larger flood plains to absorb inundation, and permitting more but smaller floods (Van Alphen 2020; Wikipedia, Room for the River; Kimmelman, 2013; McVeigh, 2014).

In effect, the Room for the River project left it to the rivers and floodplains to police themselves, absorbing overflows in expectable places.[4] One price of this was that farmers living behind the location of the original dikes became much more exposed to flooding than before. And the response to this was to move farmhouses back to artificially elevated areas safe from flooding — a twenty-first-century equivalent of the ancient terps. And this, of course, recreated the ancient poetic choreography of agency with rivers and farmers oscillating back and forth, with a timing determined by the riverflows and the agency of nature.

3. The previous examples concern poetic technologies, material artifacts: a nick in a dike, reconfigured dikes on rivers. My final example relates to the other aspect of poiesis, namely, the poetic *stance* in the world, a human disposition to practices of finding out. This is a story not so much about water as about land reclaimed from water, the Oostvaardersplassen (Zwart 2003, 123; Lorimer and Driesen, 2014; Vera, 2009; Theunissen 2019). The Oostvaardersplassen, OVP for short, is a polder north of Amsterdam reclaimed from Lake Ijssel in 1968. It was initially designated for industrial development, but this did not happen and the site was abandoned, setting in motion unexpected ecological transformations. Rare species of birds, such as spoonbills, bitterns, marsh harriers, and bearded tits began to breed in the OVP in large numbers, and other birds that had disappeared from the Netherlands returned, including greylag geese, great white egrets, white-tailed eagles, and ospreys. The geese in particular transformed the landscape of the OVP in surprising ways: "They graze the marshland vegetation . . . to such an extent that closed reedbeds have turned into open water, something which until then it was thought that only human management can achieve" (Vera 2009, 32).

We can see this phase of the history of the OVP as a classic letting go (on the human side), doing nothing. We could also see it as a simple dance of agency: the human engineers drain the land, then let it go to see what nature will do in response. And, of course, it is interesting to see it as another instance of poetic experimentation in the wild (chapter 4), interfering with nature to find out how it will react in specific circumstances. The reappearance of rare and missing birds and the transformation of the landscape are, then, striking emergent phenomena thrown up in this experiment.

Having noted earlier the importance of science in enframing water, it is interesting to note that, symmetrically, the OVP has been subject to criticism as *not* properly scientific:

> The ICMO [International Commission for the Management of the Oost-vaardersplassen] argue that SBB [Staatsbosbeheer] are not conducting a legitimate experiment. They first invoke the epistemological criteria used to evaluate secluded research, to argue that SBB is failing to comply with the fundamental Popperian requirement of seeking future falsification and requiring the full disclosure of data (Popper 1945). They suggest there has not been enough transparency in the data collection and publication to qualify this as a rigorous laboratory experiment. Furthermore, by not stating an explicit protocol for testing a hypothesis, the management regime cannot be held to account. (Lorimer and Driessen 2014, 177)

Again we have the sense of poiesis as doing without science — here, science as understood in the classic Popperian sense.

The experiment has continued. Heck cattle and Konik ponies were introduced to the OVP in the 1980s followed by red deer in the 1990s, setting in motion more emergent transformations, creating open grassland as home to large numbers of greylag and barnacle geese and wigeons, as well as lapwings and golden plovers.[5] Developments at the OVP thus continued to exemplify the poetic stance of finding out how an ecosystem will perform, the antithesis of the traditional Dutch stance of mastery and domination, a foregrounding rather than backgrounding of nonhuman agency.[6]

That is the main point I wanted to make about this example — it helps emphasize the paradigm shift from enframing to poiesis that has recently been taking place in the Netherlands (and elsewhere) and the form poiesis has taken there.

We should, though, finish the story. Developments at the OVP became a leading experiment in a global movement in "rewilding" (Lorimer and Driessen 2014; Lorimer et al. 2015; Monbiot 2013) aimed at the restoration of a prehuman ecosystem (with Heck cattle standing in for their now-extinct predecessors, aurochs, and the Konik ponies likewise standing in for extinct wild horses).[7] The experiment was controversial in the light of established expectations that the OVP would revert to forest, and the appearance over time of stretches of open water and open grassland was surprising and emergent precisely in contradicting this prediction (Vera 2009). Unfortunately, acting with nature is not necessarily nice, and another emergent effect was that numbers of ungulates died or had to be culled in harsh winters, leading to public outcries. In 2018, it was decided that the herbivore population would be kept constant and the Konik horses removed altogether. "It is clear . . . that the wilderness ideal . . . has been officially relinquished by the authorities" (Theunissen 2019, 343). The OVP experiment, in that sense, is now over, though, of course, various forms of rewilding continue around the world.

. . . .

That is as far as I can take this overview of changes in styles of water (and land) management in the Netherlands over the centuries. It serves, I think, to clarify and enrich our thinking about enframing and poiesis by adding to the field of examples. Now I want to comment on it from various angles.

First, we can assimilate the earlier studies to the broad pattern seen in this chapter. Enframing the Mississippi (chapter 2) is clearly isomorphous with the dominant Dutch style of managing rivers and the sea. Diversions in the Mississippi levees, managed retreat, the traditional style of erosion control in Japan (chapter 3) and adaptive management of the Colorado River (chapter 4) all parallel Dutch examples of poiesis. It seems reasonable to say that together these studies point to a very broad shift taking place in paradigms of water engineering from enframing to poiesis, both as engineering stance and as technology and technique.[8]

Second, the three examples discussed above all have important aspects of letting-go, recognizing and trading on the nonhuman agency of nature, and we could go a little further in reflecting on that. In some respects, letting-go can be seen as straightforwardly ameliorative, aiming to address iatrogenic problems precipitated by earlier enframing interventions — creating a nick in a dike, for example, as a step toward ecosystem restoration. But it is worth emphasizing that letting-go can also be productive in its own right, as we can see in the emergent transformations of the OVP, setting up the open-ended evolution of a new landscape.

Beyond that, a qualification is in order. The transition from enframing to poiesis in water management certainly counts as a Kuhnian paradigm shift, an inversion of gestalt, respectively backgrounding or foregrounding the agency of nature, acting on or with nature, emphasizing or relinquishing, to some extent, dualist control. But a standard reading of Thomas Kuhn (1962) evokes an image of homogeneous communities of practitioners and practices which flip entirely from one state to another: once upon a time, scientists were all Aristotelians, then they all became Newtonians, and so on. That is probably not the best reading of Kuhn (Pickering 2001b), but I want to say that it is not, in any event, how we should think of the transition here. I have listed some examples of novel poetic practices and technologies in water management, but these have by no means entirely displaced technologies of enframing. There are still plenty of well-maintained dikes, barriers, and pumps in the Netherlands (and around the world), and even the original terps were built to resist the agency of the watery environment. No doubt in the heyday of enframing we could also find examples of the stance of poiesis. Enframing and poiesis can, then, coexist in the same place and community. As discussed in the opening chapter, there is also an inevitable aspect of technical hybridity in play here: the Room for the River project, for instance, entailed plenty of conventional engineering.

The point to note is, then, that the paradigm shift we are considering is best understood not as the complete displacement of one stance by another but as a shift in emphasis or balance, another sort of gestalt switch, in which poetic practices have come relatively to the fore, standing out against a background dominated by enframing. We could think here of notions of yin and yang from traditional Chinese philosophy, with yin (the feminine principle, poiesis) and yang (the masculine principle, enframing) as two sides of the same coin and history as an oscillation of the two, with first one and then the other appearing in the foreground. In these terms, this book is an attempt to help move the pendulum in the yin direction.[9]

. . . .

Danger. From the start, I have paid attention to the dark side of enframing — the possibility that control can invite disaster. Here, centuries-long attempts to dominate rivers like the Rhine have proved to precipitate increasingly massive floods, just as in the case of the Mississippi. And in an obvious way, poetic strategies like the Room for the River project aim to attenuate this danger, seeking to substitute and live with more small floods in place of occasional massive ones.

As I have said, this attenuation of the dangers of enframing is an important reason to be interested in poetic alternatives in patterns of action. It is undoubtedly an important motivation for the paradigm shift now in progress. But the attenuation of danger is not all that is going on here, and I do not want the story of this book to revolve exclusively around it. I am interested in exploring poiesis as an interesting and unfamiliar pattern of action in its own right. The rewilding of the Oostvaardersplassen, for example, had little if anything to do with the dangers of enframing, and the same goes for the many projects and devices I have discussed in *The Cybernetic Brain* and elsewhere. I want the emphasis to be on the poetic paradigm itself.

. . . .

This is a window that shows us how the Netherlands looked thousands of years ago. —FRANS VERA, quoted in Elizabeth Kolbert, "Recall of the Wild"

I can note here as I did before that although I did not intend it, my examples of poiesis in this chapter have an aspect of going back in time and undoing what has been done. At the macro level, the story of Dutch water management is cyclical, beginning with terps and ending with them centuries later. At a more micro level, the examples often concern the restoration of environments and ecosystems, with rewilding as the most extreme example of that in its attempt to recreate prehuman ecosystems. I should therefore reemphasize, first, that I am not recommending going back in time tout court. The terps of today are not identical to ancient ones and they are embedded in very different technical, economic, and social systems. But, second, I am interested in what we can call a nonmodern stance and nonmodern technologies and techniques, and one — not the only — place to look for inspiration is obviously the premodern world.

. . . .

And God said, Let us make man in our image, after our likeness: and let them have dominion over the fish of the sea, and over the fowl of the air, and over the cattle, and over all the earth, and over every creeping thing that creepeth upon the earth. So God created man in his own image, in the image of God created he him; male and female created he them. —KING JAMES BIBLE, Book of Genesis

Science and technology...are out of control,...Christianity bears a huge burden of guilt. —LYNN WHITE JR., "The Historical Roots of Our Ecological Crisis"

My primary interest in this book is with different patterns of acting in the world, enframing, and poiesis. But these patterns of action often resonate with wider patterns of thought. In my introduction, I noted that enframing hangs together with a Cartesian dualism of people and things. In chapter 4 I discussed the relation between adaptive management and cybernetics, and I am also struck by resonances with broad religious and spiritual ontological understandings. This chapter is a good point to begin a discussion of this topic which we can continue in those that follow.

We can start with enframing and a connection to Christianity. Hub Zwart (2003, 111) notes that the move toward enframing in the Netherlands "coincided with a transition on the spiritual or ideological level ... the Christianization of the northern and western parts of the Netherlands," and many authors have emphasized this sort of connection between Christianity and the exploitation of nature. The verse from Genesis quoted at the start of this section points to a theological asymmetric dualism of man and nature — man as "the image of God" — and an associated human "dominion" over nature. Dualism predates Christianity, but we can think of this branch of Christianity as naturalizing and putting its moral force behind enframing.[10]

A dualist interpretation of Christianity and enframing thus hang together, and this might lead us to wonder about other religions and beliefs and more poetic patterns of action. In a talk titled "The Historical Roots of Our Ecological Crisis," delivered on December 26, 1966, in the early days of the US environmental movement, the historian of technology Lynn White Jr. (1967) opened up this question.[11] Having argued for a connection between medieval Christianity and enframing, he remarked that "what we do about ecology depends on our ideas of the man-nature relation," and concluded that we need to "find a new religion, or rethink our old one" (White 1967, 1206). He mentioned Zen Buddhism as "very nearly the mirror image of the Christian view" but ended up recommending an alternative interpretation of Christianity, revolving around St. Francis of Assisi as a heretical non-dualist Christian and "a patron saint for ecologists ... Francis tried to depose man from his monarchy over creation and set up a democracy of all God's creatures" (1206–7).[12]

We can think further about eastern philosophy and spirituality in later chapters, but it interests me that White's analysis again envisages going back in time and redeveloping the Franciscan path not taken, a different version of Christianity from the present mainstream.

From a different angle, there is another form of spirituality we might also think of going back to: animism. Christianity sought to eradicate animism, the pagan weather gods (Zwart 2003, 109), idolatrous sacred groves (White 1967, 1206). But perhaps there is something about these beliefs worth recovering in that they point precisely to the unpredictable agency of nature and to a world we have to get along with and cannot control — a worldview that hangs together not with enframing but with poiesis. Animism is the topic of chapter 8, but we can think briefly about it here as a counterpoint to the enframing conception of Christianity. A long way from the Netherlands, for example, the Tonga people of Zambia and Zimbabwe understood the Zambezi River to be "the home of a river god named Nyaminyami, with the head of a fish and the twisting whirlpool-like body of a snake," and to them, "the whole idea of stopping the river [by building the Kariba Dam] was absurd" (Serpell 2020, 8). And it is clear that the Tonga were very nearly right, at least in engineering terms. In 1956, construction of the dam was halted when a hundred-year flood "drowned the cofferdam that was in place for construction" (Serpell 2020, 9). The next year an even bigger flood carried away a suspension bridge and overflowed a new higher-level cofferdam. Since the dam was completed, water spilling over it has "eroded the dam's foundations and carved a pit at its base. Its plunge pool is now a 266-foot-deep crater" (Serpell 2020, 3). Nyaminyami's revenge.[13]

So, if the agency of nature is obscured by a Christian dualist ontology, it is foregrounded in animism. Animism as a form of spirituality, then, resonates with poiesis in practice. It also reflects back as a critique of enframing, as a way to remind ourselves that we are not, in fact, in control. If we saw nuclear power stations, say, as temperamental gods like Nyaminyami that can unpredictably turn against us, we would have a more accurate appraisal of how the world is than a default expectation that technologies will always work as we expect — think of nuclear disasters like Three Mile Island, Chernobyl, Fukushima (and for many more examples, see Perrow 1984). Neo-animism would introduce a much-needed element of fear into our relation with technology as enframing (Pickering 2017a, 2017b).

In parting, I should make clear that while it might be a good idea for some, my argument is not that a new religion is necessary to open up new directions for action. I want to suggest that attention to a simple ontology of emergent becoming is enough to open a space for alternatives to enframing. Conversely, however, it is also the case that a poetic ontology of unknowability can shade into a sort of minimalist mystical spirituality, though I cannot follow that line of thought further here.[14]

6 Natural Farming

The meadow with its interacting multiplicity of species is unendingly dancing.
—GREGORY BATESON, quoted by Peter Harries-Jones, *A Recursive Vision: Ecological Understanding and Gregory Bateson*

We can leave the water for the land. The topic now is farming. I first sketch in some familiar features of conventional farming as a contrast with what follows. The main topic of this chapter is a technique of so-called natural farming developed in Japan by Masanobu Fukuoka, which I see as a radical and poetic alternative to the usual form. I discuss Fukuoka's path to natural farming as an example of the poetic stance, and the mature form of natural farming as an established poetic technique. Then we can look at Fukuoka's rejection of science and his closeness to cybernetics, the difficulty of the transit from enframing to poieis, natural farming as going back in time, and the relation of natural farming to Eastern philosophy.

. . . .

Even the cultivation of the field has come under the grip of another kind of setting-in-order, which *sets upon* nature. —MARTIN HEIDEGGER, "The Question Concerning Technology"

We'd had agriculture for 7,000 years, and we'd been losing for 7,000 years—everything was turning into desert. —BILL MOLLISON, quoted in Scott London, "Permaculture: A Quiet Revolution"

Farming, especially agribusiness, the sort of industrialized farming that has multiplied and intensified since World War II, is enframing writ large. It acts on the land, animals, and plants not just to produce food but to optimize food production. In this chapter we are concerned with farming crops (rather than animals), and here I list some of the basic aspects of the process.[1]

Crop production begins with a transformation of the land. Plowing the land to break up the soil is emblematic of this sort of farming and normally a necessary preliminary to sowing seeds.[2] The historian of technology Lynn White Jr. (1967, 1205) went so far as to argue that the Western stance of enframing and domination actually originated around the seventh century in plowing, with the invention of a "new kind of plow, equipped with a vertical knife to cut the line of the furrow, a horizontal share to slice under the sod, and a moldboard to turn it over." The new plows attacked the land with "such violence" that "Man's relation to the soil was profoundly changed. Formerly man had been a part of nature; now he was the exploiter of nature." The heavy plow was an exemplary and prototypical technology of enframing that disrupted a previously more poetic relation of the human and nonhuman. And plowing was, of course, just a first step in the mechanization of farming, extending to weeding, harvesting, and beyond.[3]

Other techniques have also long been used in farming to enhance and enframe the growth of crops: the use of fertilizers to improve the soil, herbicides and pesticides to control weeds and insects, and so on. We could think, too, about inner transformations of the plants themselves: selective breeding and, more recently, genetic modification to optimize productivity and resist insects and diseases. In short, modern agriculture aims to act on the land and turn it into a food-producing machine. And various observations are in order.

First, our earlier examples of enframing have a largely negative aspect, a sort of domination by denial. Dams stop the flow of rivers; levees and dikes stop rivers overflowing their banks.[4] In contrast, the techniques just listed enhance the agency of crops in specified ways (e.g., making them grow more and faster). We could associate the former with Michel Foucault's (1979) concept of sovereign power and the latter with disciplinary power (Coppin 2003). The distinction is not important to the present study, except that it marks another axis along which science enters into enframing. The development of chemical enhancements to productivity such as fertilizers and herbicides has depended increasingly on the detour through the scientific laboratory and the chemical industry since World War II.[5]

So, it is easy to see conventional agriculture as enframing, with science as its accomplice. One last thought here goes back to the dark side of enframing and connections between domination and disaster. In previous chapters, I found myself thinking in terms of sudden breakdowns — the river that floods, levees that break, the dam that collapses. One can find similar sudden events in the world of farming. Many people warn of the vulnerability of monocrops to new diseases (like the vulnerability of humanity to the coronavirus) with the Irish potato blight of 1845 and the consequent famine as a striking historical example (Tsing 2012, 147). But there is also an argument that agriculture in itself is intrinsically a long, drawn out, rather than abrupt, environmental problem. Thomas Smith (n.d., 2) cites a long list of authors who conclude that conventional farming depends on "natural habitat destruction and expropriation, de-forestation, release of soil carbon through ploughing, leaching of soil nutrients from exposed topsoil and other impacts." As before, this dark side encourages an interest in other ways of going on.[6]

. . . .

These four principles of natural farming [no cultivation, no chemical fertilizer or prepared compost, no weeding by tillage or herbicides, and no dependence on chemicals] comply with the natural order and lead to the replenishment of nature's richness. All my fumblings have run along this line of thought. It is the heart of my method of growing vegetables, grain, and citrus. —MASANOBU FUKUOKA, *The One-Straw Revolution*

Farming as just described is paradigmatic of enframing in action and is our dominant form of agriculture today — farming as we usually think of it. But there are variant forms that look more or less different and which are growing in importance. Organic farming and permaculture both seek to minimize or eliminate the throughput of chemicals, for example.[7] Here we can explore one version of the latter that is most perspicuous for our interests, so-called natural farming as developed in mid-twentieth-century Japan by Masanobu Fukuoka (1913–2008) and set out by him in *The One-Straw Revolution* (Fukuoka 1978).[8]

First, I need to emphasize just how different natural farming is from conventional approaches. Its most striking feature is its "do-nothing" stance. In contrast to conventional farming, the land is not plowed or worked over in any way; no weeding is done and there is no sustained flooding (as is usual in rice farming); no chemicals are used, either as fertilizer or pesticide. We can go into more detail later, but the basic activities of natural farming are simply scattering seed, harvesting the crop (in this case, alternately rice and barley growing in turn in the same field in autumn and spring), and mulching the ground with the straw left over from threshing. Importantly, Fukuoka reports that using simple methods like these, it is possible to obtain crop yields comparable to if not greater than labor-intensive or chemical-dependent forms of farming.[9]

Natural farming is thus so different that when I first heard of it I was shocked. I live in the British countryside; I'm used to seeing farmers plow their land every year; and the possibility that you can farm without plowing never crossed my mind before. If I were Japanese, the nonflooding of rice fields would no doubt have shocked me just as much. I therefore want to inquire first how this radical form of agriculture arose and then what its mature form looks like in practice, emphasizing the poetic aspects of both.[10]

. . . .

We can start with Fukuoka's path to natural farming. He recalls that it began in 1938 with a spontaneous philosophical revelation that we, humanity, are *doing too much*, setting ourselves apart from nature as its masters (Fukuoka 1978, 7–12).[11] This is enframing, and Fukuoka thought that it must be possible to get out of it, to remain in or at least closer to nature and take part in the natural flow of events instead of ordering them around, to act with nature — poiesis.[12]

He set out to test this idea, working initially on his father's land, made up of one and a quarter acres of rice fields and twelve and a half acres of mandarin orange orchards (Korn 1978, xix), and what followed was a poetic and performative dance of agency that mirrors our previous examples. His first thought was simply to let his father's orchards go — to not look after them at all. This failed completely — all the trees died (Fukuoka 1978, 13). "From that point on the question, 'What is the natural pattern?' was always in my mind. In the process of arriving at the answer I wiped out another 400 trees, Finally I felt I could say with certainty, 'This is the natural pattern'" (Fukuoka 1978, 16).

Since leaving the trees alone failed, Fukuoka took a more active role, looking for ways to approach his goal of doing nothing progressively rather than in a single leap. After World War II he turned to raising crops. He immediately abandoned plowing, and to avoid herbicides, weeding, and flooding, he tried planting white clover as groundcover that would enrich the soil while inhibiting (not eradicating) weeds. This worked. Reasoning that to avoid fertilizers one had to return as much goodness to the land as possible each year, he redistributed all the straw that remained after threshing each crop. That ran into difficulties. Looking more closely, it seemed the problem was that he had thrown the mulch onto the land in clumps (Korn 2012, 8) or distributed it too carefully and geometrically (Fukuoka 1978, 48). Either way, the mulch was inhibiting the germination of the crops. And, trying it another way, he discovered that mulching was a success if he simply broadcast the straw randomly across the land (Fukuoka 1978, 48).[13]

Many more twists and turns followed. In the early days, he found it necessary to control insects with a naturally prepared insecticide (pyrethrum), but as his fields developed he found he could dispense with that too. He occasionally sprayed his trees with an emulsion of machine oil against "scale" (Fukuoka 1978, 34). On his account, sparrows were an enormous problem, pecking up the seed as quickly as he threw it across the ground. "I tried scarecrows and nets and strings of rattling cans, but nothing seemed to work very well. Or if one of these methods happened to work, its effectiveness did not last more than a year or two" (52). The solution proved to be a new chore — coating the seeds in mud pellets before sowing them (43, 51).[14]

Fukuoka found that if he let ducks wander on his fields, their droppings would help the mulch decompose more quickly. Then a new road divided his fields from the ducks' houses and Fukuoka had to take over, lightly covering the mulch with duck or chicken droppings (Fukuoka 1978, 37). At one point he rented some land from a neighbor and obtained a good crop using his usual approach, except in one corner where the "plants came up too thickly and were attacked by blast disease." When asked, his neighbor stated he put all his chicken droppings there, from which Fukuoka learned something about over-feeding his crops (37). "I once thought there would be nothing wrong with putting ashes from the fireplace onto the fields [as he did in his kitchen garden]. The result was astounding. Two or three days later the field was completely bare of spiders. The ashes had caused the strands of web to disintegrate. How many thousands of spiders fell victim to a single handful of this apparently harmless ash?" (28).

The upshot of this complex dance was the mature form of natural farming. Fukuoka had learned how to obtain reliably good yields of crops with a minimum of effort, while dispensing with otherwise definitive elements of farming such as plowing, weeding, flooding, using chemical fertilizers and pesticides, and performing much hard labor. We can examine this mature form in the next section, but first I want to comment on the trajectory to it we have just discussed.

The point I want to emphasize is, of course, its isomorphism with our short story of erosion control in chapter 3 and with the program of adaptive management of the Colorado River we reviewed in chapter 4. All these stories center on and foreground deliberately staged dances of agency — trying this, trying that, seeing what happens, reacting to that — in the way that I take to be definitional and exemplary of the stance of poiesis. Indeed, just like the traditional approach to soil erosion in chapter 3, there was little else to Fukuoka's route to natural farming besides this dance. Like the adaptive management of the Colorado, we can also see Fukuoka's stance as experimentation in the wild, performative experimentation on the thing itself — his father's orchards, then rice and other crops — finding out how they would behave in this configuration or that, with no detour through the laboratory. In fact, Fukuoka had no interest in science and the production of knowledge, a point we will return to.

....

Now we can turn to the mature form of natural farming as a more or less stable technique. Fukuoka grew fruit and vegetables as well as seed crops, but I will concentrate on the latter here for simplicity. Korn (1978, xxiii–xxiv) gives an extensive account of Fukuoka's method:

> In the fall Mr. Fukuoka sows the seeds of rice, white clover and winter grain onto the same fields and covers them with a thick layer of rice straw. The barley or rye and the clover sprout up right away; the rice seeds lie dormant until spring.
>
> While the winter grain is growing and ripening in the lower fields, the orchard hillsides become the center of activity. The citrus harvest lasts from mid-November to April.
>
> The rye and barley are harvested in May and spread to dry on the field for a week or ten days. They are then threshed, winnowed, and put into sacks for storage. All the straw is scattered unshredded across the field as mulch. Water is then held in the field for a short time during the monsoon rains in June to weaken the clover and weeds and to give the rice a chance to sprout through the ground cover. Once the field is drained, the clover recovers and spreads beneath the growing rice plants. From then until harvest, a time of heavy labor for the traditional farmer, the only jobs in Mr. Fukuoka's rice fields are those of maintaining the drainage channels and mowing the narrow walkways between the fields.
>
> The rice is harvested in October. The grain is hung to dry and then threshed. Autumn seeding is completed just as the early varieties of mandarin oranges are becoming ripe and ready for harvest.

Mr. Fukuoka harvests between 18 and 22 bushels (1,100 to 1,300 pounds) of rice per quarter acre. This yield is approximately the same as is produced by either the chemical or the traditional method in his area. The yield of his winter grain crop is often higher than that of either the traditional farmer or the chemical farmer who both use the ridge and furrow method of cultivation.

A lot is going on in this quotation, all of which points to the poetic rather than enframing character of natural farming as a technique. To begin with, there is the simple and striking "not doing" of the key elements of enframing: plowing, weeding, flooding, transplanting, using chemicals.[15] Earlier parallels here would start from my idea of letting New Orleans go (abandoning the aim to enframe the Mississippi) and include managed retreat, the Room for the River project (partially dismantling and relocating dikes), and rewilding the Oostvaardersplassen — leaving the agency of nature to do the work. Again, we can note the productive aspect of letting-go. Abstaining from conventional practices in facts precipitates, for example, a major transformation of the soil itself, as discussed below. But we should also note that Fukuoka's natural farming is more complicated than simply letting-go and entails also the sort of choreography of agency I sketched out in connection with the adaptive management of the Colorado.[16] Thus, the timing of planting and reaping the different crops is so contrived that their growing seasons overlap. In this way, each fosters the growth of the other, while — in conjunction with the planting of clover and a brief flooding of the rice paddy in June — inhibiting (not eradicating) weeds. Likewise, the alternation acts as a form of disease control: rice diseases die out during the barley phase and vice versa (Fukuoka 1978, 48). As Fukuoka put it, "If seeds are sown while the preceding crop is still ripening in the field, those seeds will germinate ahead of the weeds. Winter weeds sprout only after the rice has been harvested, but by that time the winter grain already has a head start. Summer weeds sprout right after the harvest of barley and rye, but the rice is already growing strongly" (Fukuoka 1978, 38).

The rice, barley, clover, weeds, and water thus act together in different combinations at different times of the year to achieve the desired outcome, so we have here further examples of the same kind of synchronization that characterized the Adaptive Management Program on the Colorado. And again we should remember that the farmer is coupled directly into this choreography as part of the poetic process. The seeds, obviously, do not sow themselves and the whole process is orchestrated by sowing timed to the season of the year. Mulching too requires a human farmer. And the timing of human interventions is more closely bound up with nonhuman agency than is perhaps so far apparent. Thus Fukuoka (1978, 38) outlines how "timing the seeding in such a way that there is no interval between succeeding crops gives the grain a great advantage over the weeds. Directly after the harvest if the whole field is covered with straw, the germination of weeds is stopped short." Likewise, speaking now of growing vegetables among the trees in the orchard, "the important thing is knowing the right time to plant. For spring vegetables the right time is when the winter weeds are dying back and just before the summer weeds have sprouted. For the fall sowing, seeds should be tossed out when the summer grasses are fading away and winter weeds have not yet appeared. It is best to wait for a rain which is likely to last for several days" (Fukuoka 1978, 66–67).[17]

Thus the sense that natural farming is a poetic practice very different from conventional farming. The natural farmer appears as a partner of the land, acting with nature and tuned into its rhythms and performances in a stabilized choreography of human and nonhuman agency — in the plane of practice rather than hovering above it in a position of mastery (Pickering 1995a).

. . . .

In my introduction, I mentioned some of the short- and long-term dangers associated with conventional agriculture, and it is worth making it explicit that this dark side is largely absent from natural farming. Diseases that flourish in homogeneous monocrops are impeded by a mixture of species (Tsing 2012), and, as just mentioned, the alternation of rice and barley, for example, inhibits the carry-over of diseases from one crop to the next. Other dangers of modernity center on degradation of the soil, leading to soil erosion and carbon release and requiring compensation by the use of fertilizers. Natural farming in contrast enriches the soil (with clover, mulch, etc.) and deliberately abstains from chemical fertilizer. Again, environmental damage results from the runoff of excess fertilizer, which is just absent from natural farming.[18] Whereas conventional farming exhausts and pollutes the land, natural farming feeds it.[19]

. . . .

My basic aim in this chapter is to contrast enframing and poiesis as different ways of relating to the environment, different paradigms, but it is also interesting and important to think about *transitions* between paradigms, especially, how do we get from enframing to poiesis? In my earlier examples, the transit was easy in practical terms (though not as a break from established traditions of thought and action). Adaptive management of the Colorado involved a straightforward change in the pattern of flows through the Glen Canyon Dam; rewilding of the Oostvaardersplassen involved, in the first instance, just doing nothing. But the case of natural farming is different in an interesting way. As outlined above, it took Fukuoka decades to make the transit from conventional farming to the mature form of natural farming. Why was that?

Thomas Kuhn (1962) argued that when paradigms change, the world changes too. He meant that in a metaphorical sense, but in the case of farming we can take it literally. The ground of farming, the soil itself, is different in conventional and natural farming:

> The reason that man's improved techniques seem to be necessary is that the natural balance has been so badly upset beforehand by those same techniques that the land has become dependent on them. . . . Look over at the neighbour's field. . . . The weeds have all been wiped out by herbicides and cultivation [plowing and flooding]. The soil animals and insects have been exterminated by poison. The soil has been burned clean of organic matter and micro-organisms by chemical fertilizers. In the summer you see farmers at work in the fields, wearing gas masks and long rubber gloves. These rice fields, which have been farmed continuously for over 1,500 years, have now been laid waste by the exploitive farming practices of a single generation. (Fukuoka 1978, 15, 33)

Conventional farming in effect turns the soil into a passive matrix in which fertilizer can be converted to foodstuffs, and there is a sort of *lock-in* at work here: to move away from such a regime is to invite disaster, as when Fukuoka simply stopped tending to his father's orchards. In the end, in raising crops, he had to experiment with ways to progressively transform his soil away from the traditional regime, and this is why his path to natural farming and the transition from enframing to poiesis was long and drawn out. This was dramatized recently at a much larger scale in Sri Lanka, when a ban on fertilizer imports led to a sharp drop in crop yields (Wipulasena and Mashal 2021).[20]

. . . .

The path I have followed, this natural way of farming, which strikes most people as strange, was first interpreted as a reaction against the advance and reckless development of science.... Because the world is moving with such furious energy in the opposite direction, it may appear that I have fallen behind the times.—MASANOBU FUKUOKA, *The One-Straw Revolution*

In earlier chapters, I noted that our examples of poiesis have the quality of going back in time, and I do not need to belabor this point in the case of natural farming since Fukuoka was keen to make it himself: "During the past few years the number of people interested in natural farming has grown considerably.... That which was viewed as primitive and backward is now seen to be far ahead of modern science.... A thousand years ago agriculture was practiced in Japan without plowing, and it was not until the Tokugawa Era 300–400 years ago that shallow cultivation was introduced. Deep plowing came to Japan with Western agriculture.... [I]n coping with the problems of the future the next generation [will] return to the non-cultivation method" (Fukuoka 1978, 19–20). Again we find the idea that humanity took the wrong path, in Japan with the plow four hundred years ago, and that we should rewind history and set off in a new, poetic, and nonmodern direction.[21]

....

No time, no time.... The only way to get back is to throw away the knowledge! Just become foolish like a bird or baby. —MASANOBU FUKUOKA, *Mother Earth News*

It is difficult enough to model two species growing together; to model the 100 or 200 species that most forest gardens contain is quite beyond reductionist science methods. —MARTIN CRAWFORD, *Growing a Forest Garden*

Nature as grasped by scientific knowledge is a nature which has been destroyed, it is a ghost possessing a skeleton, but no soul. —MASANOBU FUKUOKA, *The One-Straw Revolution*

Natural farming is a dance of agency leading into a repeatable choreography, and it is poetic in both these respects. And, as I mentioned earlier, natural farming is poetic also in not depending on science. Fukuoka's experimentation was always focused on the thing itself—land and crops. It took no detours through the lab and did not draw on unsituated scientific knowledge. Its foregrounding of maneuvers in the field of agency was in that sense parallel to the soil erosion example of chapter 3 and more purely nonscientific than, say, the adaptive management of the Colorado.

We could leave it there. My argument is that poiesis belongs to a different paradigm from enframing, one that foregrounds agency instead of knowledge, and natural farming certainly exemplifies that. But there is more going on in the case of natural farming and it is interesting to go further into that.

We should start by noting that Fukuoka's training and early career was actually as a scientist. He graduated from Gifu Agricultural College in plant pathology in 1935 and then worked at the Plant Inspection Division of the government's Customs Bureau inspecting plants that were coming into or leaving the country. "A believer in science, I spent most of my time peering into a microscope in the laboratory" (Fukuoka 2012, 1; see also Fukuoka 1978, 5–6). In 1938, following his "do-nothing" revelation, Fukuoka resigned his position at the Customs Bureau, but in the same year, after his first disastrous experiment with his father's trees, he left farming again and found a new job in science. In disgrace at home,

> I was offered the post of Head Researcher of Disease and Insect Control [at the Kochi Prefecture Testing Station]. I imposed upon the kindness of Kochi Prefecture for almost eight years. At the testing center I became a supervisor in the scientific agriculture division, and in research devoted myself to increasing wartime food productivity. But actually during those eight years, I was pondering the relationship between scientific and natural agriculture. Chemical agriculture, which utilizes the products of human intelligence, was reputed to be superior. The question which was always in the back of my mind was whether or not natural agriculture could stand up against modern science. When the war ended I felt a fresh breeze of freedom, and with a sigh of relief I returned to my home village to take up farming anew. (Fukuoka 1978, 14)

Fukuoka, then, was not a stranger to or enemy of science, but in the postwar period he regarded his development of natural farming as "a contradiction to the assumptions of modern science" (Fukuoka 1978, 29). This was part and parcel of his determination to remain close to nature; he felt that scientific knowledge distances us from the world (just as I have argued that science reinforces the stance of enframing). But he also developed a more substantive, ontological, argument for not relying on science, which we can examine now.

Fukuoka did not dispute that one could move plants and insects into the lab and study them, and he never disputed individual pieces of scientific knowledge. His concern was instead with the difficulty — he felt, impossibility — of completing the detour through the lab and reliably aggregating individual bits of scientific knowledge to a useful picture of the real world of farming. Running through all of Fukuoka's thinking about science one finds an image of the world as a great chain of being, a web of endlessly reverberating dances of agency, and this, according to him, is what defeats the efficacy of science in farming.[22]

We can see how this goes in his ideas about pest control.[23] He discusses, for example, contemporary attempts to protect pines from an outbreak of weevils by spraying pesticide from helicopters:

> I do not deny that this is effective in the short run, but I know there must be another way. Weevil blights, according to the latest research, are not a direct infestation, but follow upon the action of mediating nematodes. The nematodes breed within the trunk, block the transport of water and nutrients, and eventually cause the pine to wither and die. The ultimate cause, of course, is not yet clearly understood. Nematodes feed on a fungus within the tree's trunk. Why did this fungus begin to spread so prolifically within the tree? Did the fungus begin to multiply after the nematode had already appeared? Or did the nematode appear because the fungus was already present? . . . Furthermore, there is another microbe about which very little is known, which always accompanies the fungus, and a virus which is toxic to the fungus. Effect following effect in every direction, the only thing that can be said with certainty is that pine trees *are* withering in unusual numbers. (Fukuoka 1978, 39–40)

You can eradicate the weevils but this just leaves the remainder of the endless network, including nematodes, funguses, and viruses, to do their thing in the new weevil-less context, whatever that will emergently turn out to be.[24]

Fukuoka's ontological argument thus points at an endless performative regress in the great chain of nonhuman being which can frustrate our interventions, and it is interesting that he develops a parallel sociological and epistemological argument about an endless regress of scientific experts: "Methods of insect control which ignore the relationships among the insects themselves are truly useless. Research on spiders and leaf-hoppers must also consider the relation between frogs and spiders. When things have reached this point, a frog professor will also be needed. Experts on spiders and leaf-hoppers, another on rice, and another expert on water management will all have to join the gathering. Furthermore, there are four or five kinds of spiders in these fields" (Fukuoka 1978, 28).

And on top of these difficulties with assembling normal science and normal scientists, Fukuoka insists that the world contains unexplained mysteries, refractory to science: "The phenomenon of these great swarms of spiders, which appear in the rice fields in the autumn and like escape artists vanish overnight, is still not understood. No one knows where they come from, how they survive the winter, or where they go when they disappear" (Fukuoka 1978, 28).

Fukuoka's critique of science is thus both ontological and epistemological, about what the world is like and about how we know it. It is also practical inasmuch as it is a warning about, in this case, targeted technoscientific interventions in pest control. Returning to our theme of enframing and disaster, he argued that "people cannot know what the true cause of the pine blight is, nor can they know the ultimate consequences of their 'remedy.' If the situation is meddled with unknowingly, that only sows the seeds for the next great catastrophe. No, I cannot rejoice in the knowledge that immediate damage from the weevil has been reduced by chemical spraying. Using agricultural chemicals is the most inept way to deal with problems such as these, and will only lead to greater problems in the future" (Fukuoka 1978, 39–40). Again, this is not an abstract point. We could think of Rachel Carson's classic book, *Silent Spring* (1962), which contributed importantly to the birth of the environmental movement. Her argument centered on the unexpected side-effects of the use of DDT: it kills insects, certainly, but also turns out to kill populations of songbirds.[25]

How then should natural farming respond to pests if not by eliminating them? Fukuoka's basic idea was that, with a little assistance — further performative experimentation in the wild — naturally farmed land can become quite (not totally) resistant to pests, as pests and their predators come into dynamic equilibrium. The dance of agency more or less choreographs itself:

> I think that everyone knows that since the most common orchard "pests," ruby scale and horned wax scale, have natural enemies, there is no need to apply insecticide to keep them under control. At one time the insecticide Fusol was used in Japan. The natural predators [of the pests] were completely exterminated [along with the pests], and the resulting problems still survive in many prefectures. . . . Most farmers have come to realize that it is undesirable to eliminate predators because in the long run greater insect damage will result. . . . If . . . the insect communities are left to achieve their natural balance after [spraying with a solution of machine oil which is relatively harmless to the predators], the

problem will generally take care of itself. This will not work if an organic phosphorous pesticide has already been used in June or July since the predators are also killed by this chemical. (Fukuoka 1978, 59–60)

Just as the basic form of natural farming had to be found in performative experimentation, then, so did modulations of it such as pest control.[26]

. . . .

The face of nature is unknowable. Trying to capture the unknowable in theories and formalized doctrines is like trying to catch the wind in a butterfly net. —MASANOBU FUKUOKA, *The One-Straw Revolution*

In chapter 4, on the Colorado River, we explored resonances between specific poetic practices and cybernetics, and we can continue the exploration here. I do not need to repeat the preceding discussion — concerning, for example, Ross Ashby's multi-homeostat setups as models for enframing and poiesis, which illuminate natural farming just as they do adaptive management — but we can look at some of the new angles which appear in connection with natural farming.

1. At the most fundamental level, natural farming and cybernetics share an ontology, a vision of the world as unknowable, as a place that we can never master and which can always surprise us. But neither treats this vision as a recipe for passivity. Both fields are concerned with the question of how to act in such a world.

2. Fukuoka shared with the cyberneticians an image of the world as a multiplicity of reciprocally interacting agents (trees, weevils, fungi, viruses), and just as Fukuoka was interested in the ways that predators control pests, Ross Ashby, one of the founders of British cybernetics, developed an abstract mathematical analysis of control in multielement systems. He noted that variations in one element can be important in controlling variations in others, and his discussion of "stabilizing the stabilizer" showed how fixing one element of a system could lead, in fact, to wilder instabilities in others. He showed, for example, how the imposition of price controls in postwar Britain could lead to new instabilities in the overall economy (Ashby 1945; Pickering 2010, 147). Fukuoka's argument against eradicating pests in farming is a special case of Ashby's general point.

I think that the functioning of such hierarchies may be compared with the business of trying to back a truck to which one or more trailers are attached. . . . Each added segment denotes a drastic decrease in the amount of control....When we consider the problem of controlling a second trailer, the threshold for jackknifing is drastically reduced, and control becomes, therefore, almost negligible. As I see it, the world is made up of a very complex network (rather than a chain) of such entities which have this sort of relation to each other, but with this difference, that many of the entities have their own supplies of energy and perhaps even their own ideas of where they would like to go. —GREGORY BATESON, *Steps to an Ecology of Mind*

Another of the founders of cybernetics, Gregory Bateson, became an important figure in the US environmental movement in the late 1960s. As in this quotation, he also conceptualized the environment as a multiplicity of interacting entities, and he used that image as an argument against illusions of control implicit in, for example, the large-scale use of pesticides.[27] Like Fukuoka, Bateson thus thought that it is perhaps impossible to predict the outcome of our interventions on the environment.

As I said at the beginning, one of my reasons for embarking on this book was the hope of following Bateson's concern with the environment further, but I became frustrated (as I often have with Bateson) by his emphasis on nondualistic *thinking* and his lack of attention to *action*.[28] As in his work on cybernetic psychiatry, he offered no clue to specifically poetic forms of environmental engagement. In *The Cybernetic Brain* (2010, chapter 5) I argued that R. D. Laing at Kingsley Hall in London showed how Bateson's psychiatry could translate into a radically novel "anti-psychiatric" practice. Similarly, we can see Fukuoka's natural farming complementing Bateson by providing a lived instantiation of a cybernetic relationship with the land.

3. There are interesting and revealing parallels between natural farming and the biological computing project of two cyberneticians, Stafford Beer and Gordon Pask, from the late 1950s and early 1960s (Pickering 2009b). The ambition of biological computing (never quite realized) was to use naturally occurring lively systems (such as a pond ecosystem) as computing elements. This is in contrast, of course, with conventional computing, which starts by rearranging the structure of matter, silicon chips, right down to the atomic level. In analogy, natural farming operates on already organically lively soil, in contrast to conventional farming, which respecifies the soil at the molecular level as a neutral matrix for fertilizers, herbicides, and so on. So biological computing and natural farming are both imaginative and poetic examples of *working with* nature as found. Conventional computing and conventional farming are examples of the enframing of matter itself, working very deeply on the inner constitution of nature to put it at our command.

4. Both natural and conventional farms are orderly complex systems existing far from equilibrium, in the sense that left untended each would run down to a disordered and unproductive state, and it is interesting to think about cybernetic accounts of self-organization in this respect (von Foerster 2014; Prigogine and Stengers 1984; Gleick 1987). The standard story of the appearance of order in systems far from equilibrium is that it requires transits of energy and matter through those systems. In a simple example, the appearance of convection cells in a fluid depends on a heat supply from below and heat loss from above. This obviously fits the description of conventional farming with its incoming and outgoing cascade of chemicals.[29] It does not fit the description of natural farming with its absence of outside inputs. And this perhaps helps us appreciate the singularity of natural farming as a little island of self-organized stability isolated from all else — including the material flows that plug conventional farming into webs of technoscience.

5. A last axis of resonance between natural farming and cybernetics is a shared interest in Eastern philosophy and spirituality. Fukuoka was Japanese, so it is no great surprise to find Eastern threads to his thought and practice, and we can explore these further in the next section. In a way, it does seem surprising to find the East running through the work and lives of the twentieth-century Englishmen I wrote about in *Cybernetic Brain*. It certainly surprised me when I first came across it. Stafford Beer, for example, taught tantric yoga. But in the end, it is not so surprising. Working on nature, enframing, is a dualist and characteristically Western stance of domination. Conversely, working with nature is a poetic, nondualist stance which removes humanity from the center. In note 1 of the opening chapter, I listed briefly some contemporary, largely academic, nondualist perspectives, but for many of us the obvious place to look for nondualist inspiration is eastward — Buddhism, for example, as dismantling the Cartesian ego at the heart of the dualist worldview. Hence, I think, this final overlap (Pickering 2011).

Before leaving cybernetics, I should ask: So what? What have we learned from this digression? Fukuoka would probably have said, "Nothing." Detours and digressions just take us away from thing itself, growing food, and should be avoided. But from my perspective, these links to cybernetics help widen the frame, assimilating natural farming to the very different examples I set out in *The Cybernetic Brain* as well as to the other examples in this book, and thus filling out the space of an overall poetic paradigm as another way to be.

．．．．．．．．．．．．．

The spirit of technology without theoretical science seems...to be found within Taoist philosophy itself. —JOSEPH NEEDHAM, *Science and Civilisation in China*

Putting "doing nothing" into practice is the one thing the farmer should strive to accomplish. Lao Tzu...would certainly practice natural farming. —MASANOBU FU-KUOKA, *The One-Straw Revolution*

In the previous chapter, I discussed interconnections between enframing and a dualist strain of Christianity, and between poiesis and animism. Similarly, resonances between natural farming, poiesis, and nondualist strains in Eastern philosophy and spirituality invite further exploration here, as Tom Smith (n.d.) has shown.[30] I am certainly no expert in this area, and, of course, "the East" has fostered a multiplicity of different traditions (and likewise the traditions that come to mind are not confined to any geographical area). But still, I find some ideas I associate with Eastern philosophy both suggestive and illuminating in thinking about poiesis (and, by contrast, enframing), and what follows is a speculative exploration of these. My aim is to pick out some new resources for thinking constructively about natural farming and about poiesis more broadly. I begin with some simple general observations. Then we can turn to two key concepts connecting Eastern thought specifically to questions of agency and performance: *shi* and *wu wei*.[31]

First, if mainstream thought in the West has always tended to an asymmetric dualism of people and things, Eastern traditions have pointed instead to a symmetric engagement, nicely caught up in a small Chinese jade sculpture (figure 6.1). The main elements of the sculpture are towering mountains and clouds, but if you look closely you can see small figures and buildings nestled into the landscape. The overall vision, then, is of humanity as just a part of the world, certainly not its dominating presence, but perhaps a decorative addition. We could say that this sculpture *puts us in our place*. And natural farming (like our other examples of poiesis) exemplifies this: the farmer as tuning him- or herself into a nature which he or she cannot control but can, at best, get along with.

Second, if Western thought privileges knowledge and especially science, Eastern thought often tends to undo that privilege. Meditation, for example, as a classic technology of the self, seeks to clear the mind of all preconceptions and routinized associations. Zen koans, riddles without solutions (the sound of one hand clapping), "exhaust...the intellectual process so that a clearer view of reality would reveal itself.... By discarding all learned doctrines and knowledge, a person is able to achieve real unity with the Tao"

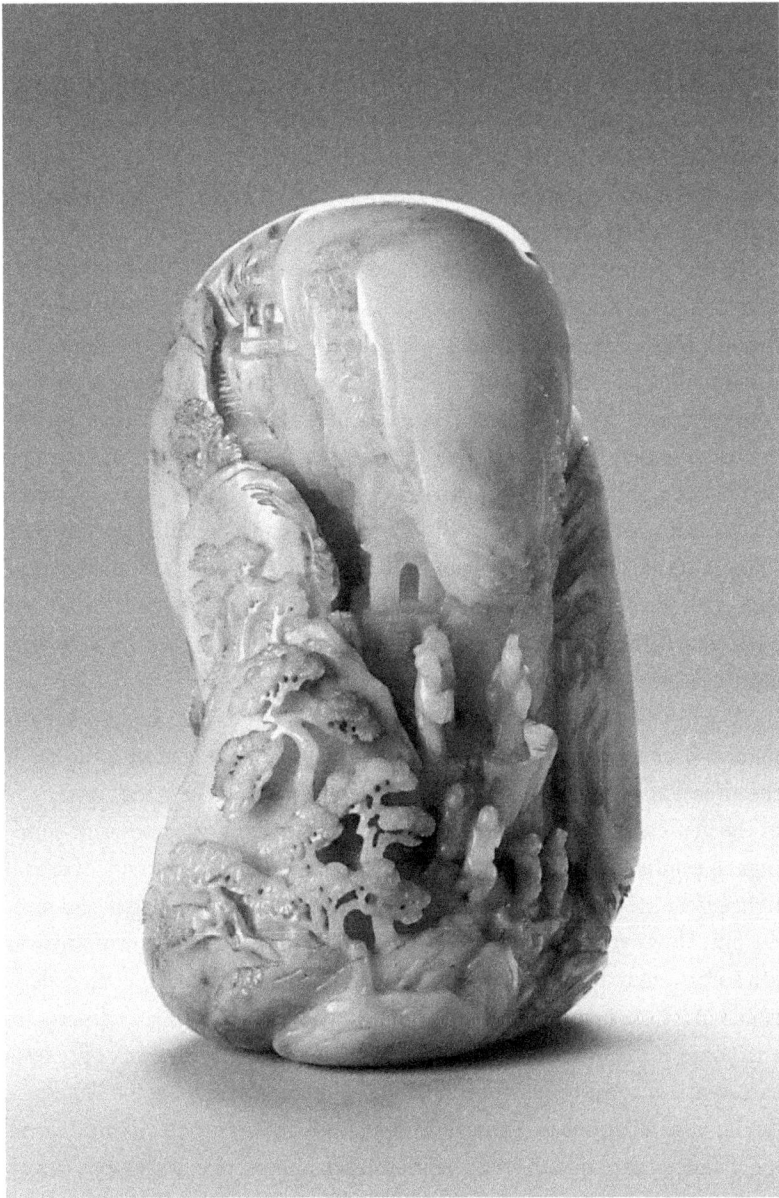

FIGURE 6.1 Jade mountain and landscape. China, Qing Dynasty, early eighteenth century. Durham University Oriental Museum.

(Juniper 2003, 22, 18). "For Daoism, 'selection and rejection' are precisely what should be abandoned. A natural life, embodied in the sage, is to 'learn not to learn' (Laozi 2006, chapter 64) or to 'exterminate learning' (Laozi 2006, chap. 20)" (Yu 2008, 8).[32] Again, the relation to natural farming with its aversion to the detour through science is clear. In his rejection of science, Fukuoka refers repeatedly to the Buddhist critique of the "discriminating mind" (e.g., Fukuoka 1978, 124).[33]

In very general terms, then, mainstream Eastern philosophies map nicely onto Fukuoka's worldview, and we can note that, in the quote above, Fukuoka himself aligned natural farming with Laozi — the author of the *Daodejing*, the foundational work of Daoism.[34] But now I want to clarify some further connections to the East which are not specific to Fukuoka, but which interest me because they bring out the performative aspects of both natural farming and Eastern philosophy. They can help us grasp what is going on in poiesis.

We can start with the concept of *shi*, which François Jullien (1999) says is ubiquitous in Chinese thought and which he translates as the "propensity of things" — the disposition of specific things to act in specific ways.[35] From our perspective, the key feature of *shi* is that it indeed refers directly to performance, action, doing things — hence my interest here.

In Chinese thought *shi* has immediate implications for human performance: human action is proper and effective if it is in tune with the *shi* of the situation in which it takes place. Jullien's first example is the classic one of the general who exploits the *shi* of his human and nonhuman surroundings to render the enemy powerless and thus avoid the necessity of actual fighting. From one angle, of course, just what *shi* is remains foggy and mysterious. How are we to know the *shi* we are dealing with? The usual answer is a sort of inner noncognitive mastery (or tuning-in) based on long experience.[36] But here I am tempted to deviate somewhat from Jullien's discussion and come at the question differently. I find it suggestive to think of poiesis as a stance of actively *finding out* and taking advantage of what the *shi* of particular constellations is. Thus, in an obvious way, en route to natural farming, Fukuoka explored the propensities to act of trees, rice, barley, rye, clover, sparrows, pests, weeds, water, and the land, finally arriving at a complicated circular choreography of *shi*. And, of course, the same is true of our earlier examples as well.[37] On this admittedly speculative understanding, then, just as enframing acts out a dualist ontology of man's dominion over nature, so natural farming and poiesis more broadly can be seen to act out a decentered ontology of *shi*.

All very simple. The concept of *shi* thus interpreted helps us get to grips with the development and practice of natural farming and its performative center of gravity very directly. Conversely, one virtue of this account of natural farming and our other examples is that they can help Westerners grasp a possible meaning of *shi* and what it means to act in accordance with it—which is what I have been referring to as poiesis. The complicated thing to understand from that perspective would not be the meaning of *shi* but all the twists and turns and detours back and forth of enframing, including modern scientific knowledge and instrumentation and the trip through the lab. Which is as much as to say that starting from a *shi* ontology not only illuminates the stance of poiesis but also serves to bring our own dominant patterns of thinking and acting into focus; to problematize what we usually take for granted, mastery and domination, science and engineering; to make them strange.

The second concept that interests me here is *wu wei* — often translated as *not doing* — a key concept of Eastern philosophies. In his extensive exploration of *wu wei* in both Confucian and Daoist traditions, Edward Slingerland (2003) conceptualizes *wu wei* as "effortless action," again understood as skilled performance but now in a very generalized sense as the ability to make the appropriate response in any and all situations.[38] But here I want to follow Sandra Wawrytko's writing (2005) on "Daoist ecology" in thinking through specific cases — "concrete ecological examples of how those concepts [including *wu wei*] apply in practice" (80).[39] Discussions of *wu wei* typically note that it means more than not doing in a literal sense, but I am tempted to suggest that it can indeed include this. Fukuoka often described his route to natural farming that way, as "let's try not doing this, then not doing that . . ." (plowing, weeding, fertilizing, etc.). We can also think here of other comparable letting-gos: Fukuoka and his father's orchard, the city of New Orleans, farmers in the Exe Valley, the Rhine and Room for the River, the rewilding of the Oostvaardersplassen. "Effortless action" in this sense would mean just that: letting the agency of nature simply shine through.

But we can expand the picture by thinking of natural farming as a poetic technique, in which these literal not-doings are themselves entwined with more interesting maneuvers aiming to latch onto and gracefully insert the farmer into the agency of nature in complex choreographies: the overlapping of rice and winter crops as both weed and disease control, the careful timing of sowing, pest control as finding a balance with predators, and so on. My suggestion is, then, that natural farming can be seen from this angle as a rich exemplification of *wu wei*. Enframing is the dominating contrast-class: it is our way — forcing the other into submission.

To get at the contrast between enframing and poiesis, I have talked about shifts of gestalt, in which different elements come to the fore or recede into the background. The idea is that poetic practice foregrounds nonhuman agency and dances of human and nonhuman agency, while enframing backgrounds them. Now we could contrast Eastern and Western philosophy in the same way. Western philosophy — in the dualist form that has come down to us via Descartes from the ancient Greeks — backgrounds performance and nonhuman agency, while important threads of Eastern philosophy foreground them. This is why Eastern philosophies in particular help us grasp poetic practices like natural farming. It also explains why it is hard for us to grasp the existence of alternatives to enframing: it entails a gestalt shift — a revolution! — away from dualism and the natural ontological attitude of the West.

Finally, we can pursue this line of thought a little further. The great historian of science and civilization in China, Joseph Needham (1971, 234–35), discussed two different approaches to engineering to be found in ancient China. One, which maps onto my concept of poiesis, he called Daoist hydraulics, for which "*wu wei* was the best watchword." Opposing this was a Confucian approach to engineering which favored building dikes to contain rivers, in just the same way as the US Army Corps of Engineers has attempted to control the Mississippi (Needham 1971, 249). His exemplar (249–50) of Daoist engineering is the ancient Dujiangyan control structure on the Min River in China, which acts in different ways depending on the quantities of water flowing through it. When the river is low, it directs the flow mainly into a very extensive irrigation system; when levels are high most of the flow goes into a bypass to avoid flooding. The Daoist and poetic aspect of this is that the physical structure of the Dujiangyan dam itself enters into a complex relation with the agency of the river, going with the flow, just as natural farming enters into and choreographs the agency of soil, crops, and so on. Contrast this with conventional dams like the Glen Canyon Dam in chapter 4, which enframes and dictates terms to the Colorado. Built around 256 BC, Dujiangyan is still in use today and became a World Heritage Site in 2000. Just as Heidegger invoked a silver chalice as his exemplar of poetic technology, we could think instead of Dujiangyan.[40]

7 A Choreography of Fire

An interest in the dark side of enframing sensitizes us to disaster. Bridges, dams, and levees have always failed, but I have noticed this much more since I started working on this book. What used to seem like unfortunate background noise to world history now strikes me as a pattern: mastery sometimes undoes itself catastrophically. The topic of this chapter is fire. I am interested in this because so many disastrous wildfires raged all around the world while I was writing (2020–21) — in Australia, California and the American West, Portugal, the Amazon, Siberia, Turkey, Greece, and the Mediterranean. I am especially interested in fire because besides disasters, one can identify a very different, nondisastrous, productive, and poetic relationship with fire, which is the concern of this chapter. As usual, I begin with some brief remarks on fire and enframing to serve as a contrast with what follows. I focus throughout on Australia since a poetic relationship with fire there has been documented in impressive detail in Bill Gammage's book *The Biggest Estate on Earth: How Aborigines Made Australia* (2011), which is my principal source of information in this chapter. Having sketched out key features of Aboriginal fire, I turn to the contrast with an alternative modern and scientific approach to controlled burns, echoing Masanobu Fukuoka on the cumbersome form of the latter. The chapter concludes with brief discussions of Aboriginal animism, danger, skill, going-back, and social relations.

. . . .

Fire in Australia has a rich and complicated history (Pyne 2020), but we only need an outline. Prior to the coming of the Europeans, the native Aborigines burnt the land a lot and early European settlers found this "horrific" (Gammage 2011, 158). Back home, fire was the enemy, and this has been the predominant reaction ever since. We could think of fire as akin to an invasive species like the eels of chapter 1, and likewise the object of attempts at eradication. Much postsettlement Australian history thus consists of attempts to enframe fire, in the brute sense of stamping it out.[1] This enmity has long been institutionalized in the form of organized fire-fighting forces (Franklin 2008, 33–34), and, as usual, science and technology have been increasingly drawn into the fight, including planes and helicopters to drop water on fires, as well as drones, sensors, AI, and predictive computer modeling (Sullivan 2020).

We need not go further into the modern fire-regime, except to note, first, its dark side. The aversion to fire translates into an accumulation of highly combustible material on the ground. Franklin (2008) emphasizes that eucalyptus trees, for example, shed flammable bark and branches, so that forested areas become disasters waiting to happen. Wildfires are possibly rarer now than hitherto but more intense in proportion to the aversion to fire. In Wolfgang Schivelbusch's terms (1986), the "falling height" of wildfire increases with attempts to enframe it. Like levees and Mississippi floods, fire control conjures up its own nightmares.[2] "After the removal of the Aborigines, bush fires became more explosive and dangerous and a litany of major disasters were experienced: Black Thursday, 1851 (Victoria), Black Sunday, 1926 (East Melbourne), Black Friday, 1939 (widespread in eastern Australia), Black Tuesday, 1967 (Tasmania), and the Canberra Holocaust, 2003" (Franklin 2008, 32). Now we can add the even bigger fires of 2019–20. On April 21, 2020, the *New York Times* reported that, at last, "on March 2, for the first time in 240 days, not a single bush fire burned in the state of New South Wales. The state's Rural Fire Service declared the worst fire season in history, during which 25 people in NSW were killed, officially over. In those eight months, 6 percent, or 13.6 million acres, of the state that a third of Australians call home had been incinerated" (Sullivan 2020).

Second, we should note that one approach to enframing wildfire is to fight fire with fire. Controlled burning of limited areas aims to decrease the amount of combustible material scattered across the land and thus reduce the potential for major wildfires.[3] Stephen Pyne (2020) records that there have long been arguments for and against this strategy between various groups including foresters and settlers. In fact, in the early 1950s, "Australian forestry . . . did something that no other industrialised country did: it adopted controlled burning, not fire suppression, as the basis for the protection of its wildlands. . . . Systematic burning, informed by science and restrained by bureaucracy, could lessen damages, improve chances for fire control, reconcile Australians with their environment, and not least project a distinctive Australian identity to the world" (Pyne 2020, 70, 72). But this policy in turn was strongly criticized by environmentalists, and "state foresters found their capacity to fight fire crippled, and their understanding of how to manage fire, challenged" (78). We will discuss scientific controlled burns later. In any event, these have never been enough and wildfires continue to rage.

....

In reaction to recent fires, many authors have noted that before European settlement of Australia, the Aborigines handled fire differently, and still do, given the chance. Often this is represented in largely ameliorative terms, as simply preventing something worse. Following the customary usage, Adrian Franklin (2008, 23, 29) and Lisa Asplen (2008, 168) both refer to Aboriginal fires as "cleansing" or "cleaning up" the land — the removal of material that would otherwise feed larger blazes.[4] This is to see Aboriginal fire from the present and as only addressing present problems. But much escapes this description.

Speaking of the deep past, James Scott (2017, 38–39) outlines a much richer appreciation of fire:

> Our ancestors could not have failed to notice how natural wildfires transformed the landscape: how they cleared vegetation and encouraged a host of quick-colonizing grasses and shrubs, many bearing desired seeds, berries, fruits, and nuts. They could also not have failed to notice that a fire drove fleeing game from its path, exposed hidden burrows and nests of small game, and, most important, later stimulated the browse and mushrooms that attracted grazing prey. Native North Americans employed fire to sculpt landscapes favored by elk, deer, hare, porcupine, ruffed grouse, turkey, and quail, all of which they hunted. The game they subsequently bagged represented a kind of *harvesting* of prey animals they had deliberately assembled by carefully creating a habitat they would find enticing. Quite apart from being the designers of hunting grounds — veritable game parks — early humans used fire to hunt large game. The evidence suggests that long before the bow and arrow appeared, roughly twenty thousand years ago, hominids were using fire to drive herds of animals off precipices and to drive elephants into bogs where, immobilized, they could be more easily killed. Fire was the key to humankind's growing sway over the natural world — a species monopoly and trump card, worldwide.

I have quoted this passage at length because it conveys the diversity of productive uses of fire, from modulating the growth of trees and grasses to luring and hunting game. Fire was historically more like a sophisticated form of farming than any simple cleansing. Scott refers to precolonial fire in New England (Cronon 1983), but by far the most detailed and extensive account of pre-European fire is Gammage's *The Biggest Estate on Earth: How Aborigines Made Australia* (2011), to which we can now turn.[5]

....

Fire was a life study. Seasons vary, rain is erratic, plants have life cycles, fire has long and short term effects, people differ on what to favour. How each species responded to fire had to be set against deciding which to locate where.—BILL GAMMAGE, *The Biggest Estate on Earth*

Gammage is interested in the period around 1788 when the first English settlement in Australia, then a penal colony, was established. His interest is to reconstruct from contemporary accounts (from visitors and settlers) how the people, as he calls them, the Aborigines, managed the land in this period.

The most striking feature of Australia to early visitors was how well kept it appeared, often resembling the parkland of the English nobility: open stretches of grass with a scattering of large trees (Gammage 2011, chapter 1). The entire continent shared this worked-over quality, which is why Gammage describes it as resembling a single enormous estate. The other striking feature was the fondness of the people for fire; visitors remarked on the number of fires they saw burning in different places at different times. Gammage makes the connection and explores at great length just how the people used fire to reshape their environment. We do not need to go into great detail, but I can try to summarize some of Gammage's key findings. What I need to stress is that fire was not a blanket technology of enframing to be imposed from above independently of its context. Instead, it was tuned poetically into the *shi* of the land, the propensities of things, including plants, animals, geography, and the weather.[6]

The material technology of Aboriginal fire was and is very simple. Traditionally, firesticks, burning and smoldering brands, were used to ignite fires; nowadays, disposable cigarette lighters are used (Verran 2002). The techniques and practices of fire, in contrast, have been subtle and varied. We can start with the mechanics of fire and then turn to its uses in acting on plants and animals.

An Arnhem man explained, "you sing the country before you burn it. In your mind you see the fire, you know where it is going, and you know where it will stop. Only then do you light the fire." (Gammage 2011, *The Biggest Estate on Earth*)

In 1788, most fires were not accidental. They were deliberately started and managed by the people. The aim was to target specific areas with burns that would last for finite lengths of time. Wind direction was important here, carrying fire in a predictable path across the land. Natural features could then function as firebreaks, setting a limit to burns. "Fire managers burnt precisely. . . . They aimed fires at breaks or patches" (Gammage 2011, 165). "Hot fires were 'aimed at stone, sand or earlier breaks'" (Gammage 2011, 173).

Temperatures and weather conditions were also important in determining the duration of fires. "People timed most (not all) fires to go out at night: overnight fires could confess loss of control. To decide what day, even what hour, to burn, managers took account of wind, humidity, aspect, target plants and animals, and fuel loads. They burnt on hot, windy days if it suited, but wind was commonly an autumn or winter ally. It aimed fire; dew, rain, frost or snow restrained it" (Gammage 2011, 169). "Later fires were generally downwind of protected sites, and lit when the grass was dewy in early morning or late afternoon, depending on how far the fire must travel. In the cold time cold air and ground moisture stifle fires soon after dusk, so most were lit about three hours before. Burning stopped when the winds came about late August" (56). "Many newcomers saw fire before rain. . . . Walter Smith Purula, a southern Arrernte elder, made clear to Kimber how important rain was, and how much else must be considered" (170). "As the Wet ended, north Queenslanders burnt a succession of fires to make a mosaic of recovery stages, stopping in the hot months of the late Dry, when 'fire climb up over the mountains . . . it might kill some yam and all that you know, kill the trees, too hot'" (163).[7] Interestingly, other natural phenomena could function as proxies for the weather: "They watch the white ants — when they start carrying their eggs out of the creek and put them on a high place, then they know it's going to rain. They start burning again" (170).[8]

Aboriginal fire was thus a poetic technique, tuned into the local *shi* — the propensities to act, the agency — of wind and weather, the land, plants and animals, even insects, coupling them together in a choreography of agency, just as the engineers in chapter 4 coupled waterflows through the Glen Canyon Dam to floods on the Paria and Little Colorado Rivers and Fukuoka coupled together the propensities of plants, birds, insects, and the land.

Now we can turn from the mechanics of fire to its uses, starting with its effects on vegetation and moving on to animals. Fire cleared grasses and affected trees without necessarily destroying either. But it did more than that. It encouraged the regrowth of different species in different ways, and could thus be used to reconfigure vegetation in a targeted fashion. One important parameter here was the frequency of burns. For instance, "To convert eucalypts to grass people had to let fuel build up so fires could run, but burn often enough to kill seedlings, and maintain this over many generations until the old trees died. Burning most eucalypts every 2–4 years would in time make grassland, while burning a little less often would let some saplings survive and create open woodland" (Gammage 2011, 13–14). Aboriginal fire thus played with the varying agencies of different plants and trees. We could think here of engineers exploring how frequent artificial floods on the Colorado should be.

Besides frequencies, intensities of fire were important in structuring its effect. Different species of vegetation respond positively to different degrees of fire, so modulating the intensity of a blaze was another tool for controlling subsequent growth. "Particular animal and plant communities needed and got very precise fire timings and intensity. Trickling cool fires and clearing hot fires were timed to keep species in balance. On dry Sydney sandstone people burnt cool fires in spring, but hot fires in early summer to open hard seeds and pods or germinate legumes" (Gammage 2011, 167). And more complex mediations tuned both the intensity and frequency of fire to foster the growth of different species with different propensities, the kind of balancing act in spaces of agency that Fukuoka entered into in controlling pests, for example.[9]

> Burning every 2–4 years promotes perennial grasslands. In 1788 these were common, which means they got with unbroken regularity the fires they needed. They also carried annuals, bulbs and tubers killed by hot fire, but needing ash to thrive, and cool fires every 2–3 years to open the perennial canopy. No random bushfire could strike that balance, or let such unlike partners flourish so widely. All have declined since 1788. Spinifex country supports no food plants until it is burnt, when plants like Desert Raisin appear and fruit prolifically. Fruit production then drops annually until in about 5–8 years, depending on the rain, Spinifex has again smothered the plants. Of twelve food plants in the Centre, five need fire, three tolerate it, and four are killed by it. All twelve flourished in 1788, so people managed them with different but adjacent fire regimes over many centuries. (Gammage 2011, 14)

Fire is a technique for managing vegetation, which can be an end in itself, but also a way of managing animals. On the one hand, different fire regimes were important to the very survival of different species:[10] "Mallee fowl survive only small, patchy fires, frequency depending on terrain, so people burnt often enough to limit fuel, but carefully enough to let the birds flourish. Gliders and possums like frequent fire, but rat kangaroos need casuarinas burnt about every seven years, a native mouse needs heath burnt every 8–10, mainland tammar wallabies need dense melaleuca burnt very hot every 25–30. In the Centre, 'When the little emus are on the ground [about August] you do not burn'" (Gammage 2011, 168).

Very importantly, kangaroos like lush grasslands, so fostering patches of grass is a way of encouraging them to gather there — ready to be hunted and eaten.[11] Likewise, many animals and birds choose to live in the margins of forested areas, so fire was used to maintain not just grasslands and forests but, especially, the boundary zone between them. And Gammage goes further, examining complex "templates" of land which combined a variety of animal and vegetable resources maintained through complex fire regimes. Paintings by Joseph Lycett (ca. 1820), for example, show templates consisting of grass corridors between thick forest and portray how kangaroos could be hunted by driving them along the corridors into ambush (Gammage 2011, 91–93). Gammage's argument in the end is that at the time of European settlement Australia was a mosaic of such templates.

Finally, we can think of timing from another angle. Hunting tends to exhaust the land: once you have eaten most of the kangaroos, hunting obeys a law of diminishing returns. So mobility was also integral to Aboriginal practice, which used fire to move templates progressively into new territory while leaving other parts behind. A sort of Daoist not-doing reappears here from earlier chapters: just leaving plants and animals alone to recover.[12]

I have not conveyed the richness of the Aboriginal fire-regime that Gammage documents, but from our point of view the material discussed is enough to capture some of its key features. The overall point I want to bring home is that Aboriginal fire was by no means a dualist technology of enframing; it was not a fixed and unsituated solution to the problem of the land. Instead, it was a flexible and responsive assemblage of poetic techniques threaded into the agency and propensities of the land and the weather, vegetation, animals, birds, and insects.[13] Fire tied the people into nature — and vice versa. Aboriginal fire in its complex interweaving of human and nonhuman agencies is, in fact, the most sophisticated and weightiest of our examples of poiesis in action. Now I want to comment on it from various angles.

The key point I want to stress about poiesis is its performative skeleton which directly entangles human and nonhuman agency, but, as before, we can now think about the cognitive aspects of Aboriginal fire. Threaded through it, of course, is all sorts of down-to-earth Indigenous knowledge, as sketched out, concerning the timing, control, and specific effects of fires of different intensities. This heuristic knowledge itself reflects back to foreground acting-with and the nondualist poetic couplings that are our focus. In the other direction, it both derives and feeds into a broader ontological understanding discussed in the next section. Here, I want to think further about science and fire.

The basic and most important point is the obvious one: science had no place at all in Aboriginal fire. In terraforming a continent, no scientific detours away from and back to the land took place. Fire prior to European settlement was indeed doing without science, an organized complex choreography of human and nonhuman agency. Beyond that, however, one might be tempted to argue that although Aboriginal fire was, as a matter of historical fact, independent of science, now it might be possible to do it better, so to speak, with the aid of science. This would be, after all, the conventional story of how science has come to replace Indigenous knowledge and traditional forms of understanding and action.[14] Against that view, in the previous chapter I quoted Fukuoka exclaiming "No time!" in a discussion of science and agriculture, an expression of his idea that science opens up endless regresses of entities and agents, regresses that stand in the way of action. Once one enters into these webs, solid ground is lost and uncertainty multiplies.[15] Better, according to Fukuoka, to stick with the thing itself: performative experimentation in the wild and the poetic techniques that emerge from that. And we can see how that goes here by drawing on Helen Verran's (2002) juxtaposition of Aboriginal and scientific fire, which centers on a workshop in Northern Australia in 1996, led by Aborigines and intended to introduce scientists to their firing methods.

The first striking feature of this encounter is that Verran can describe Aboriginal firing practices very simply in just a few sentences. One day, leaving the scientists behind, in fact, a couple of Aborigines jumped in a car and set a line of fires (Verran 2002, 740). The next day, scientists were present when "a small party of men [walked] around the loop of a creek, lighting fires as they went" (Verran 2002, 742). And that's basically it. Verran records that place names and associated social groups were very important in Aboriginal explanations of this process, but her account says nothing about functional reasons for the specific fires (along the lines of Gammage's analysis of re-freshing the grass, say). "Despite their best efforts to be polite and respectful, the scientists' dismay sometimes showed. . . . [They] were able to recognize neither valid generalizing about habitats, nor justifiable strategies of burning them" (Verran 741, 742).[16]

Verran juxtaposes this with a discussion of the process of scientific "pre-scribed burning." Her exposition comprises two steps, both of which prove to be exceptionally complicated. On the one hand, she draws on an instructional science video to explore the practical production of scientific knowledge of fire and plants. The video first discusses the role of fire in Australian ecologies and the use of burning to reduce the chance of larger fires, and then goes on: "However using fire to manage the age and health of a vegetation type is more difficult. Before a programme can be developed, the rangers need to know exactly what plants are present in the area, their flowering and germination requirements, how long they usually live for, and the number of different plant species changes over time. . . . To determine exactly what sort of vegetation they are dealing with, and to help predict what is likely to happen if the area is burnt, a small sample area is often investigated first — a field-site is established" (Verran 2002, 747, quoting from the video).

The reference to a need to know "exactly" points already to an indefinite project of classification and measurement. The video then turns to extended processes of measurement, entailing staking out meter-square quadrants, tagging them so they can be reidentified after a burn, using rulers to determine positions within quadrants, making painstaking expert identifications of plants within quadrants, and taking photographs from permanently fixed posts. And this happens not once for each quadrant but repeatedly. "Resampling the quadrants will continue for many years so that plants' response to fire can be followed" (Verran 2002, 747, quoting the video).[17] Here, then, we glimpse a methodological regress, paralleling the epistemological one of identifying species, that can continue as long as one is willing to lay down quadrants and repeat measurements.

This is not nearly the end of the story. "Scientists collect data and often develop a model to reveal general characteristics of the habitat, and yet even armed with all these data it is not easy to predict the outcome of any particular burning programme. The events occurring in an area after it is burnt have a great deal of influence on the new vegetation: temperatures . . . not enough rain . . . too much rain . . . competition for water . . . soil nutrients . . . and living space. . . . Seedlings are engaged in a struggle for resources" (Verran 2002, 747–48, quoting from the video). Scientific exploration of fire and vegetation thus has the same quality that Fukuoka ascribed to the science of insect control in agriculture, a never-ending trip through a ramifying web.

So far, I have been summarizing Verran's account of the science of burning. The second component of her story concerns how to *use* this knowledge in a "prescribed burn." To give a flavor of this, she reproduces a textbook description of the protocol for a prescribed burn. This consists of more than a page of detailed instructions set out under twelve headings, running from "1. Initial planning" to "12. Surveillance," and including sections like:

5   Division of tasks. Typically, the operation is divided up — different groups have different duties and work in different sectors. One team may be dropping incendiaries from a helicopter while a ground crew (supported by a firetanker) sets fires elsewhere at precise points along the established control lines. The operation may also be sub-divided into time stages. The organization that initiates the overall plan will probably be responsible for mopping up after the burn-off for the following days until it is declared safe.

6   The lighting-up plan. This involves deciding exactly where and in what order fires will be lit. The correct sequence matters; it determines what burns first, and in what direction fires can move. Fires can be volatile when moving up steep slopes so these areas are often lit from above. The fires then burn downwards to control lines (where possible) at the base of hills, or into pre-burnt areas. Humidity also affects the behaviour of fires, so "time-of-day" lighting-up can help prevent hot burns in sensitive areas, especially if incendiary capsules dropped from a helicopter are precisely placed. This is very useful in getting a "black edge" for rain forest gullies and stream-side vegetation communities, but one has to calculate precisely. The same technique is often used in remote areas for "steering" fire. (Verran 2002, 744–46)

In contrast with the simplicity of Verran's description of the Aboriginal approach, here at the level of method we again find a quasi-endless list of instructions for a whole company of actors. It would be so difficult and time-consuming to follow the protocol for a prescribed burn that it is hard to imagine any such burns actually taking place.[18]

I have quoted Verran on controlled burning at length to show just how detailed and lengthy the processes of scientific fire are, as a way of making the contrast I want to emphasize: on the one hand is Aboriginal fire, quickly describable as a performance in and with nature; on the other, the more or less endless quality of the scientific detour through knowledge, away from and back to nature, that Fukuoka pointed to in the previous chapter — "No time!" And we could recall that the former works better than the latter. Aboriginal people kept their continent in good shape until European settlement, while the scientific settlers have recently overseen the worst fires since they first arrived. In this instance, then, far from an irrevocable advance on Indigenous techniques, modern technoscience is a cumbersome substitute.

. . . .

Country was not property. If anything it owned. —BILL GAMMAGE, *The Biggest Estate on Earth*

In chapter 6, on farming, I discussed the resonances between natural farming and Daoism, and parallel resonances are evident in Aboriginal relations to the land. Aboriginal religion and ontology are usually discussed under the heading of the Dreaming, a constellation of beliefs, songs, rituals, and practices in which I am very far from expert, but some points which are important to us are clear enough.

First, the Dreaming assumes an animist ontology. Plants, animals, and the earth itself are seen as home to lively spirits. Second, this ontology is symmetric and decentered. Aborigines saw themselves as lively agents within this lively world. At this level, then, they shared the nondualist worldview on which this book is predicated.[19] Third, ontology does not in itself imply specific patterns of practice. One can always go against the ontological grain; animism does not in itself obviate a stance of enframing. But the Aborigines went with the grain, seeking to act with rather than on nature, and thus exemplified the poetic form of life that is my focus here. Their fire techniques brought their spirituality down to earth as worldly practice.

I can take this line of thought a little further. Gammage (2011, 124) explains that the Dreaming required Aborigines to "leave the world as you found it — not better or worse, for God judges that, but the same." But interestingly, as we have seen,

> This does not mean leaving things untouched. On the contrary, the cycles of life and season change constantly, and a manager's duty is to shepherd land and creatures safely through these changes. Eddies they may be, but they are part of the Dreaming and must be cared for. This might require dramatic or spasmodic change (burning forests, culling eels, banning or

restricting a food), and it certainly demands active intervention in the landscape. Ancestors do this still, obeying the Law and seeking balance and continuity. Humans should do no less. Land care is the main purpose of life. (131)

We are back to not-doing, *wu wei* in Daoist terms, as, in fact, an active and dynamic process, as in Fukuoka's natural farming — ontological choreography, orchestrating human and nonhuman agency to maintain the overall state of a system.[20]

. . . .

Danger. Throughout the book I have emphasized the dangerous dark side of enframing, and not much needs to be added here. Wildfires have become only more widespread and dangerous while I have been writing this book. On the other hand, while one cannot deny a risk of sparking wildfires inherent in Aboriginal practices, the risk of disasters would seem to be much reduced. These days, controlled burning is recognized around the world at least as a potential antidote to catastrophic fire.

Skill. In chapter 3, I sought to distinguish my conception of poiesis from concepts like skill and *metis*.[21] There, I emphasized that a poetic stance entails staging dances of agency. Here we can think further about poiesis as an established technique. There is no doubt that Aborigines have been skilled practitioners of fire for centuries, and I have no wish to deny that. But in describing Aboriginal fire as poiesis I am trying to draw attention to something else. The key point for me is that skill is a human-centered concept. Skill in the use of fire is then something that belongs to individual Aborigines, something they have and carry around with them, just like I was once pretty skilled at table tennis and bar football. Poiesis, in contrast, is a relational concept, a posthumanist one, spanning the human-nonhuman divide. Thus I have emphasized that Aborigines act so as to release, channel, and take advantage of nonhuman agency — letting the wind carry fire along and the rain extinguish it, modulating the growth of specific plants, animals, and insects, and so on. Poiesis is positioning and attuning one's acts so as to integrate them with an active nature. It belongs to a decentered human-nonhuman assemblage, not to any individual human — and this is the difference from skill that I have in mind.

At the same time, it easy enough to see a relation between poiesis and skill. In earlier chapters, I described the stance of poiesis as performative experimentation in the wild — staging dances of agency, trying this, trying that, finding out what happens, responding to that. And this is, of course, a recipe for the acquisition of skill: skills are learned through this catalog of successes and failures. That's exactly how I learned to play darts. Needless to say, I have no historical information on how the Aborigines became skilled in fire, but it is safe to guess their skills accumulated in just such a poetic process.[22]

One last point in this connection. The reason I want to fight off references to skill is that skill is such a familiar concept in both academic and everyday discourse. We think we already know what it means and it tends to draw interpretations into its gravitational field. Against that, I am trying to point here to poiesis as something we are not used to thinking about and that remains difficult to see — a key element of an unfamiliar gestalt, a different paradigm that I am trying to pull out of the shadows of enframing.

Going back. As before, this chapter invites us to contemplate going back in time and starting again on a new trajectory, now in respect of fire. Gammage's referent is the year of first settlement of Australia, 1788, while earlier in the chapter Scott pushed similar fiery techniques back twenty thousand years. Given our present increasingly calamitous relation to fire, this rewinding of history again seems like a good idea, at least in principle. In practice, it would be difficult if not impossible. I am not in a position to think this through in any detail, so let me just cite Gammage's conclusion instead: "There is no return to 1788. Non-Aborigines are too many, too centralised, too stratified, too comfortable, too conservative, too successful, too ignorant." But if this seems too final, he is speaking, I think, about a blanket reinstatement of Aboriginal practices, and he is more optimistic about partial returns. He continues, "Yet across the shattered centuries 1788 can still teach, and some have begun to learn. Tree corridors replicate belts, wetlands are being restored, reserves and sanctuaries declared. Aborigines sometimes have more say in fire management, and whenever a city burns more people accept control fires, though these remain too few" (Gammage 2011, 321; see also Russell 2020). Perhaps, then, these ancient practices might be a sketch of another future for Australia — and other places.[23]

....

Finally, I can just mention another angle on enframing. British settlers in Australia were immersed in enframing *social relations*. Back home, social relations were hierarchical: the upper classes and the aristocracy were masters of the common people. And, of course, there was a widely held belief in a hierarchy of races, with white Europeans at the top and everyone else below them — "primitives" waiting to be dominated, guided, inducted into Christianity. Aboriginal social relations were not like that. As far as class was concerned: "Europe's social gradations are not apt to 1788 Australia, but Aborigines were more akin to Europe's gentry than to its peasantry. They commanded no-one, but they had land, sought knowledge, had much time for religion and recreation, and usually lived comfortably in parks they made" (Gammage 2011, 310–11). At the same time, "the most catastrophic idea was the notion of race. In 1788 Europeans were beginning to assume a hierarchy of races, with themselves at the top. Almost all Australians could have no concept of race until there was more than one, after invasion. The difference crippled all the negotiations people attempted in 1788, rejected their compromises, and mocked what efforts they made to 'be like us'" (Gammage 2011, 312). European and Aboriginal social relations thus stood in stark contrast to one another, and in a striking parallel somehow corresponded to different stances in relation to the land, enframing and poiesis, respectively.

What should we make of this? It would be a mistake to assume a Durkheimian or Marxist causality leading directly from relations between social groups to relations with the land. But this correspondence helps us understand why enframing came so easily to Europeans (and still does) and not to Aborigines. We could say that for each group, techniques of land management, social relations, and religion and spirituality hung together, reinforced and interactively stabilized one another as elements of quite different broad paradigms of thought and action.[24] Aboriginal social relations, then, can stand as another dimension of the poetic paradigm I am tracing out, and another candidate, social this time, for going back and rewinding history.

8 Spirits

A sorcerer seeks to act rather than to talk and to this effect he gets a new description of the world—a new description where talking is not that important, and where new acts have new reflections. —CARLOS CASTANEDA, *Tales of Power*

Our topic is acting with nature, and in this last substantive chapter, I want to go further in thinking about what sort of nature we are acting with — and also who we are. The key concept throughout has been agency, and our examples so far have featured ordinary everyday instances of human and nonhuman agency: engineers build dams, rivers flood, crops grow, and so on. We could argue about definitions of "agency" (Pickering 2023) but no one would doubt that people and things do act in such ways. This is good in getting the argument across, but it does not fully escape the modern image of the world and the distribution of powers. I could say that the previous chapters undermine modern dualism *from within*: even if we accept conventional understandings of the sorts of things humans and nonhumans can do, still we have been exploring the reciprocal couplings and tunings that undermine the dualist Great Divide, as Bruno Latour called it.

To broaden the picture, I find it helps to think about other, unconventional and nonmodern, understandings of agency, and about animism in particular — the idea that the world is populated by gods and spirits.[1] I have found myself referring to animism several times in previous chapters simply as a way to dramatize the liveliness of the nonhuman world. I even suggested at the end of chapter 5 that animism might be a useful way of reminding ourselves of the instability of technologies like dams and nuclear power stations. As a last topic I now want to go to extremes and explore what would be involved in taking animism and its attributions of agency more seriously. We do not have to, but it is instructive to try. My text here is Davi Kopenawa's *The Falling Sky: Words of a Yanomami Shaman* (Kopenawa and Albert 2013).[2] The

Yanomami are animists, natives of the Amazon rainforest in Venezuela and Brazil. Kopenawa himself is the shaman in question, and, alongside more mundane matters, his book discusses at length the process of becoming a shaman and the powers it entails.

. . . .

White people... do not dream as we do. They sleep a lot but only dream of themselves.
—DAVI KOPENAWA, *The Falling Sky*

We can start with the distinctive world the shaman inhabits. It is not like our everyday world or even the everyday world of the Yanomami. Kopenawa tells us that it is populated by the spirits that he calls *xapiri*. These are autonomous agents — willful, powerful nonhuman beings that the shaman can see and communicate and negotiate with. Approached properly, they entertain, teach, protect, cure, and act on behalf of the shaman. Approached improperly, they can kill.[3]

Animism is thus, in our terms, immediately and directly poetic: Yanomami shamans recognize, foreground, trade on, and act with the liveliness and unpredictability of their world. Indeed, the world of the shaman is very much livelier than anything that figures in previous chapters, composed of volitional and temperamental entities with magical (as we would say) powers. That is the sense in which my earlier references to animism aimed to stress the liveliness of the world, and in which the animist sense of agency in fact goes far beyond modern understandings of nonhuman liveliness.

To get clearer on the singularity of animism, it is instructive to think once more about science. In previous chapters, I have tried to show how science helps sustain the stance of enframing while often being tangential at most to poiesis. Here the relation is even more polarized. Not only is science irrelevant to the shaman; science denies the shaman's world any real existence. On the standard account, a definitive accomplishment of science has been to transcend animism and put it decisively behind us. We used to be animists but science shows we were wrong — and this on multiple counts: (1) Nonhuman agency is not willful; to think otherwise is to commit the sin of anthropomorphism; (2) Gods and spirits are not intersubjectively accessible — only shamans can see and interact with them, therefore they cannot be part of a consensual scientific cosmos; (3) The powers ascribed to spirits like *xapiri* are inexplicable on the basis of current science and are therefore illusory.

Those of us who have grown up in the secular West find it hard to argue with science here. Nineteenth-century Anglo-American spiritualism seems like a bit of a joke, like alchemy and witchcraft before it.[4] But we can press on. The key point to note is that the spirits are not omnipresent and continuously available even to the shaman. They are at hand only when the shaman is in an altered state, the achievement of which entails rigorous self-discipline that includes quiet, no women, a specific diet, and, most importantly, consumption of a vegetable drug, *yakoana*.[5] Only as the upshot of this process can the shaman come into the presence of the *xapiri* on specific occasions.

What can we say about this? We have seen that the shaman's world entails a nonstandard form of nonhuman agency — autonomous and willful entities with strange powers, the *xapiri*. Now we can see that there is another side to this, a form of human agency also very different from the modern everyday picture. As shaman, Kopenawa can do things that in their everyday state other humans cannot — see, hear, and interact with the *xapiri*. We can think of the shamanic rituals and drugs as technologies of the self in Michel Foucault's sense (1988), which help elicit complementary nonstandard human and nonhuman performances. So, the picture becomes more complex. Now animism needs to be seen as not simply a belief about the world but as an assemblage of spirits, practices, and powers and a specific kind of person that goes with them.[6]

Of course, science rejects this whole assemblage. Not only is the animist world view inherently impossible from a scientific perspective, but animism in practice hinges on a scientifically undesirable transformation of the shaman's self. The intersubjective world that science inhabits depends on sober observers in their everyday state, so that the shaman is the paradigmatic scientifically unreliable witness.[7] Again, the scientific position seems compelling, but again we can press on.

We can reverse the logic. In the science I know best, elementary particle physics (Pickering 1984a), it is not the case that everyone can see elementary particles — protons, neutrons, electrons, mesons, resonances, quarks, leptons, intermediate vector bosons. No one can see them directly and only a very few can, so to speak, access them at all. Many years of very sophisticated and arduous education, training, and research practice, up to and beyond a PhD, are required to conjure them up. Powerful and enormously expensive equipment is also required, running from particle accelerators and colliders through detectors like bubble chambers and spark detectors, up to the massive electronic arrays found today at CERN, the European Organization for

Nuclear Research. Very sophisticated mathematics and enormous computing power are also required to interpret the performance of these machines as evidence of particle properties. Enormous diversified and coordinated teams of trained physicists and engineers are needed, too, in recent generations of experiment.

In short, there is a symmetry here. Only very specially prepared people with very special equipment can enter into the presence of elementary particles, and the same goes for shamans and spirits. The shaman's special preparation and equipment — principally *yakoana* — is, in fact, less extensive and less exclusive than the physicist's. We could say that we have two different paradigms here in Kuhn's sense — two bodies of training and equipment — which bring forth different worlds, those of spirits and their strange powers, or of quarks, strings, and black holes and their strange powers. And from this angle I want to emphasize that we can take both paradigms seriously (Pickering 2017a; Vasantkumar 2022). If we were to focus solely on representational *knowledge*, we would, I think, have to come down, as we usually do, on the side of science. But if we think in terms of actions and performances, science could be "true" without shamanism being "false" and vice versa. We would simply have to see them as two different ways of standing in the world, one a stance of enframing, the other of revealing and poiesis.[8]

And, of course, having seen this symmetry we can also point to an obvious contrast. Both paradigms in a sense make the world dual: physicists encounter particles that exist independently of them; shamans face spirits that also exist independently of them. But the difference is between the asymmetric dualism on which modernity is founded and the symmetric dualism of animism. Nonhuman agency is squeezed out of the physicist's world. Only entities that are in principle lawlike and predictable are permitted entry, while nonhuman agency comes to the fore in the shaman's world: it is crucial to the shaman that the *xapiri* are willful and artful. Again we find a gestalt switch. The modern sciences background the agency of the world while shamanism and animism foreground it — my classic distinction between enframing and poiesis.[9]

Where does this get us? Partly, we are back where we started, with a contrast between two paradigms. One of them, science, conjures up the world as machine-like, predictable and controllable, and thus sets it up for enframing. Throughout, I have been seeking to challenge this image by pointing to the liveliness of even the everyday world of rivers, farms, and fire, and I have been looking at examples of the unfamiliar, nonmodern, poetic practices that foreground that liveliness. But now we have extended the image of the poetic paradigm to an extreme in seeing how it can encompass a much wider vision of the liveliness of the world to include willful spirits with strange and scientifically inexplicable powers. At the same time, this extension points to a reassessment of human agency as including shamanic powers that are not part of our everyday world, complementary to the strange powers of the shaman's world.

I do not insist that we take shamanism seriously. I have no experience of it, and nothing in this book depends on it. But I find that trying to take shamanism seriously stretches my imagination of what people and things can do and enriches my understanding of poiesis and human and nonhuman agency and their intertwining. We do not need to be imprisoned by modern understandings of what there is in the world. I also do not want to suggest that animism is the only sort of extension to our thinking on agency that we can contemplate.[10] But, for my part, I find that this excursion reflects back on us — denizens of the modern scientific world. It helps me see how strange *we* are; how much we have to modify and form ourselves and our made world to bring forth our asymmetrically dualist cosmos. It puts science in its place and points to other ways of standing in the world.

. . . .

In Antiquity, every tree, every spring, every stream, every hill had its own *genius loci*, its guardian spirit. These spirits were accessible to men, but were very unlike men; centaurs, fauns, and mermaids show their ambivalence. Before one cut a tree, mined a mountain, or dammed a brook, it was important to placate the spirit in charge of that particular situation, and keep it placated. By destroying pagan animism, Christianity made it possible to exploit nature in a mood of indifference to the feelings of natural objects. —LYNN WHITE JR., "The Historical Roots of Our Ecological Crisis"

Kopenawa, like many others who are actively resisting the onslaught of extractive industries in remote forests, may in many ways be more "advanced" in his understanding of the planetary crisis than an academic in a tranquil Western university town. —AMITAV GHOSH, *The Nutmeg's Curse*

To conclude, we can return briefly to two topics. One is *danger*. The shamanic world is by no means devoid of danger, not least, as Kopenawa tells it, to the shaman — the mortal danger of engaging with the *xapiri*.[11] But this is an individual danger, knowingly and willingly undertaken. In contrast, we have seen all along the potential for large-scale and sometimes planet-wide disasters that accompany enframing as its unintended but inevitable dark side.

The Yanomami themselves have an analysis of this contrast, which one can capture by replacing with *xapiri* the centaurs, fauns, and mermaids in the quotation above from Lynn White. On the Yanomami account, the *xapiri* maintain the health of the Amazonian rainforest, so that they are the others that shamans and the general population have to "placate and keep placated" in taking care of the forest.[12] From this angle, the greatest danger resides in *not* placating the spirits, in which case they will depart and the sky will fall.[13]

Conversely, in their asymmetric dualism, "white people . . . only dream of themselves" (Kopenawa and Albert 2013, 318). They lack any nonhuman other with whom they need to struggle and negotiate, so that they "lose all restraint" (318), they act "without measure" (382). Nothing resists the modern, Western stance of enframing, rendering the Amazon standing reserve for human exploitation, with results that are evident as fires, deforestation, and global warming.[14]

My second topic is *going back*. Inasmuch as science is said to have definitively superseded animism, my interest in poiesis and acting with nature has again led us back, now into the prescientific past. I am primarily interested in animism, as I said, as a way of stretching our conceptions of agency, but consistency requires that I note, at least, that animism is another practical thread which might be worth picking up again in the present. As in previous chapters, the argument is not that the animist paradigm should entirely displace the scientific one. But I have sketched out a way to take both animism and science seriously from a performative perspective. Seen from that angle, coexistence is entirely possible, and, as in earlier chapters, the hope is to push the yin-yang balance of modernity toward the poetic yin pole; to show that, and how, we can be interested and even involved in animism without contradiction. To repeat, simply reflecting on animism reminds us that we are not, in fact, in control.

Conclusion

Poiesis

As M.B.S. [Mohammed bin Salman, Crown Prince of Saudi Arabia] conjures this brave new world—no journey will take more than 20 minutes! zero carbon emissions!—you get the sense that his chutzpah is nothing short of metaphysical. He appears to believe that nature itself is at his command. —ROBERT WORTH, "The Dark Reality Behind Saudi Arabia's Utopian Dreams"

We have been looking at poiesis as an important and interesting but relatively unfamiliar pattern of action — acting with nature and the environment. This last chapter reviews the overall picture, picks out some key terms in capital letters, and discusses some topics that deserve further attention, concerning paradigms, engineering, politics, and, finally, education.

To start with **POIESIS** itself, I understand it as a **PATTERN OF ACTION** that recognizes and finds a place for nonhuman **AGENCY**. Agency, in turn, is **EMERGENT** — we never know what people and things will do next. So poiesis has the character of **PERFORMATIVE EXPERIMENTATION** — trying this and that, to find out what will happen and **ACTING WITH** that. In its simplest form, this can amount to **LETTING GO** — abandoning existing control structures to let the agency of nature shine through, as I have put it. Letting go can be simply ameliorative, as in the managed retreat of human settlements, but importantly it can have a constructive aspect, too, as an opening to the actions of nature, like absorbing floodwaters as in the Room for the River project, or becoming something new, as in rewilding: poiesis as **IATROGENESIS**. In its more elaborate form, poetic action entails deliberately staging **DANCES OF AGENCY**, human and nonhuman — **FINDING OUT** how the world will act and then reacting to that, and so on, back and forth. In two of our examples — the adaptive management of the Colorado ecosystem and Fukuoka's natural farming — these dances settled down as regularized **TECHNIQUES**, **CHOREOGRAPHIES OF AGENCY**, in which human and nonhuman agency are flexibly geared into and coordinated with one another. Similar choreographies are the topics of the chapters on fire and spirits (chapters 7 and 8, respectively).

To bring out the specificity of poiesis by contrast, I have also included brief examples and discussions of **ENFRAMING** as its much more familiar counterpart — our usual way of going on. The **STANCE** of enframing is one of the mastery and domination of nature, backgrounding — suppressing, domesticating, ignoring — the agency of nature and the emergent phenomena that accompany it. Enframing **TECHNOLOGIES** perform likewise, imposing themselves on the world as fixed solutions, **ACTING ON** the environment, not with it.

I have tried throughout to bring out the **PERFORMATIVE** quality of poiesis and enframing, as patterns of action — doing things in the world. But it is important to recognize that they have **COGNITIVE** aspects too, which also serve to highlight the specificity of poiesis. Thus **SCIENCE** as an organized body of finished knowledge is often central to enframing as we have seen, especially in backgrounding emergence, working as a shortcut to the future, going around the findings-out of poiesis — and, indeed, veiling the performative aspects of enframing itself. In contrast, science was largely absent from or tangential to most of our examples of poiesis, which themselves thematize and foreground not knowledge but performance: **DOING WITHOUT SCIENCE**, as I put it. Some science did figure especially as tentative modeling (rather than as finished knowledge) in the adaptive management of the Colorado, but even there we could say that the center of gravity of poiesis remained in performance not cognition.

A counterpart to science in some of our examples of poiesis was **INDIGENOUS KNOWLEDGE** — of fire in Australia and spirits in Amazonia. The key observation here is that while science presumes an asymmetric dualist **ONTOLOGY** that hangs together with enframing, the Indigenous knowledge in question instead foregrounds nonhuman agency and patterns of acting with it. Beyond my own nondualist ontology coming from science studies — the basic symmetric picture of human and nonhuman agency — I have also discussed other nonmodern ontologies as illuminating the specificity and strangeness of poiesis, including **ANIMISM** and traditional Chinese philosophy, on the one hand, and **CYBERNETICS**, on the other.

....

SO WHAT? I keep coming back to this question. Why should we be interested in poiesis? First, because it shows that we have a choice in how we go on in the world. For me, at least, this counts as a discovery. In 1985, when I first read Martin Heidegger's ([1954] 1977) famous essay "The Question Concerning Technology," I could not understand why he was going on so much about enframing. It had simply never occurred to me that it was possible to act otherwise, and I suspect this is true for many people. It was only more than a decade later that I started to appreciate the possibility of poiesis (Pickering [2000] 2023). One important aim of this book, then, is to foreground poiesis as an alternative pattern that denaturalizes enframing. Enframing is not given to us by nature or some supreme deity; we could try acting differently if we wanted to.

Why should we want to? I can think of several reasons. As I said in the opening chapter, the world is full of **WICKED PROBLEMS** that in their exceeding complexity science can find no purchase on, and poiesis can help us here by offering a measured and constructive way forward, staging dances of agency, finding out how we stand with respect to wicked systems. The Colorado River ecosystem discussed in chapter 4 is a nice example — the adaptive management program is a way of successfully feeling our way forward in a situation far too complex to see our way through.

From another angle, we could think about the **DANGERS** that come with enframing. On the one hand, in treating nature as a docile servant — "standing-reserve" (Heidegger [1954] 1977, 225) — enframing in effect cuts us off from the world; nothing new comes back to us from our interactions with it. Poiesis instead *is* an intimate, productive, performative relation with the world. It puts us **IN OUR PLACE**, as caught up in a world of becoming, not standing above it. On the other hand, enframing has disaster as its **DARK SIDE**. We can try to ignore, repress, domesticate the agency of nature with more or less success, but emergence will always burst out somewhere, sometimes catastrophically as we have seen. Nuclear power stations melt down; carbon dioxide produced as a sort of performative excess from the use of fossil fuels feeds into global warming; and so on. And the **FALLING HEIGHT** of these disasters increases in proportion to our mastery. Poiesis, in contrast, serves to avoid or at least attenuate such dangers — think, for example, of Aboriginal fire techniques, or the Room for the River project as an alternative to massive floods exacerbated by dikes and hard defenses. It would require another essay to explore this point more systematically, but it seems clear that while poetic experimentation with the agency of nature cannot be immune to risk, it is less prone to disastrous outcomes than attempts to pin nature down.

Having mentioned global warming, we can continue with the **ANTHRO-POCENE**, the state of affairs in which humanity has taken over as a prime mover in the evolution of the planet. The Anthropocene is the ultimate reductio ad absurdum of enframing. Global warming and all the other dark sides of the Anthropocene show us that (a) we are locked in a symmetric performative embrace with the world, which (b) we do not in fact control. Our actions have unintended consequences that can come back to harm us, and increasingly so. Of course, I have no magical cure for the ills of the Anthropocene. We seem to need a very quick fix, which, if it is to be found at all, will probably remain in the space of enframing. Continue to depend on gigawatts of cheap electricity, but now derived from solar, wind, and nuclear power. Stop eating meat, but otherwise farm much as before but more efficiently. Maybe geo-engineer the reflectance of the atmosphere.[1] I have no inspiration to offer for enframing-style solutions, but it seems to me that such solutions amount to rearranging the famous deck chairs on the Titanic — possibly necessary, but short-term, desperate, and brutal and themselves carrying all the usual dangers of enframing.

Conversely, the poetic projects we have looked at directly address Anthropocenic problems — poetic approaches to water engineering and flood control, adaptive ecosystem management, natural farming, Aboriginal fire techniques, animism as respect for the forest. And more generally, poiesis is a way of chipping away at the Anthropocene from below, undermining it by acting differently. The aim of poiesis is not to decouple ourselves from nature — which could never be done — but to couple us into it in new and more symmetric ways that can variously sideline, adapt to, and counteract the Anthropocene's painful and damaging aspects.

. . . .

PARADIGMS. Thomas Kuhn (1962) adopted the word *paradigm* as an insightful way of catching up great swathes of the history of science. In this book I have tried to suggest that our examples of different ways of engaging the environment are instances of two very broad paradigms, of enframing and poiesis, acting on or with the world. As Kuhn used the word, *paradigm* has proved hard to define, but I can be more specific here. The paradigms in question are patterns of action and thought distinguished by ontology or worldview: an asymmetric dualism of people and things as the mark of enframing; a world of endless becoming as the mark of poiesis. My examples explore the ways in which patterns of action speak to those ontologies, and in which the ontologies reciprocally illuminate the patterns. Paradigm talk might be regarded as dated in the history of science, but it seems to me a good way of bringing out especially how poiesis differs from our customary modes of enframing.

Having said that, various points can be clarified. First, a paradigm is not a fixed entity. It is something that grows and adapts open-endedly — Kuhn spoke of scientific practice as the business of articulating a paradigm. In the present context, that means two things. From one angle, the examples we have looked at are not fixed things. Adaptive management or natural farming continue to evolve. My examples are better seen as sketches of another future, seeds from which further poetic practices can grow. From another angle, my list of examples is by no means intended to be exhaustive. No doubt one could find many more examples of poiesis in our relations with the environment, and I have also suggested that poiesis extends far beyond that into all the fields touched by cybernetics, from brain science, psychiatry, and robotics to management and the arts (Pickering 2010).

One reason for speaking of paradigms here is thus to sweep up my diverse examples of poiesis as instances of the same sort of thing. My feeling is that while adaptive management, natural farming, Aboriginal fire, and so on are interesting in their own right, they carry much more weight as instances of a common stance in the world. In this respect, my contribution is to make these cross-connections — which sadly the practitioners themselves do not make — visible, in the hope of making the pattern available as something that might grow into the future, as a model for other ways to be. Specialization is the enemy here; I am trying to open up possibilities by joining the dots and breaking down some of the fragmentation of modernity.

Second, paradigm talk can take us in directions we do not want to go. Kuhn (1962) introduced the term to think about "scientific revolutions" — big changes in which some overall description of the world is entirely replaced by another: relativity and quantum mechanics taking over from classical mechanics, for example. The sense of radical change and difference is important to me, but the question of replacement needs thought. On one side, in some areas, one can imagine poiesis replacing enframing entirely, at least in principle. Masanobu Fukuoka claimed that his version of natural farming is as productive as contemporary technoscientific agriculture, so perhaps poetic approaches could replace enframing entirely as far as crops and vegetables are concerned (while remembering the practical difficulties entailed in this shift, discussed in chapter 6). In principle, it seems that anything, certainly including poetic Aboriginal techniques, would be better than the usual enframing approaches to fire (again remembering the practical impediments mentioned in chapter 7).

But if in some situations it is possible to imagine poiesis replacing enframing, at the other extreme, there are plenty of cases where, as far as I can make out, at present there is, as Margaret Thatcher liked to say, no alternative. The modern world is, for example, dependent on construction materials like steel and concrete, and it is hard to think of poetic routes to their production and use. And something similar can be said of science. I believe it is worth contemplating the ways in which finished science plays into projects of enframing, but that is certainly not a reason to get rid of science. It might be enough to say that I am glad to have been vaccinated several times against the coronavirus even if, and just because, it does enframe the entity in question.

These thoughts lead us into the **YIN** and **YANG** of poiesis and enframing. The suggestion of this book is not a total replacement of ways of acting on nature with ways of acting with it. As I said in chapter 6, on paradigms of water engineering, my hope instead is to **SHIFT THE BALANCE** of how we go on in the world, away from enframing and toward poiesis. I want simply to show that interesting and significant alternatives to enframing exist, both as things already in the world and as models for future practice.

Thirdly, along the same lines, we should think about **HYBRIDITY**. Yin and yang are not discrete opposites; they are intertwined (Watts 1957). As discussed further in the opening chapter, enframing calls forth poiesis in practice — as local adaptations of plans, maintenance of machines, responses to failures, and so on; poiesis in turn depends on elements of enframing — controlling waterflows in experimental floods, for example. So instances of each paradigm invoke the other. But this does not mean they are the same. They differ as **PRACTICAL GESTALTS**. As we have seen, enframing is action organized around mastery and acting-on; poiesis foregrounds finding-out and acting-with.

....

Having got this far, it seems appropriate to mention a couple of topics that I have not discussed here that readers might be interested in. This should, at least, help clarify what the book is and is not about.

Bad engineering. I have noted the many dark sides of enframing, and a standard response to these is to blame the engineers. When dams and levees and bridges fail, we ask what has gone wrong and tend to blame the engineers who designed and built them. I certainly share this tendency, and no doubt in many instances it is right to do so. So, why have I not mentioned this before? First, because I have nothing original to say about the human failings of engineers; I can leave that to others, better placed than I am. And second, because in this book, I have been after something else, something deeper, in a different space and not so well understood: the ontological fact that the emergent aspects of enframing projects are always liable to burst through somewhere, however well-engineered they are. We cannot see the future; there are no perfect engineers. "Mistakes" are often only visible in retrospect, tracing back from specific failings. And even this tracing back can lead not to the solid ground the engineers should have been standing on but into a swamp. Many years ago, I explored a scientific critique of quark-search experiments in particle physics, which turned out to lead not to certainties but into arguments about obscure electrical properties of materials about which no one had any definitive knowledge (Pickering 1981). Charles Perrow (1984) likewise shows how complicated it can be, for example, to trace back the origins of 'normal accidents' at nuclear power stations.

Moreover, many unwanted and unintended consequences of our actions have nothing to do with considerations integral to engineering design and construction. Global warming, for instance, is not a consequence of the bad design of car engines or oil-fired power stations. The engines and power stations work fine (usually). Until recently, no one saw global warming coming. I can remember when it was nuclear winter that we feared. The performative excess of enframing can always surprise us.

So, there is more going on in the dark side of modernity than bad engineering, and that something more is what has concerned us here.

Politics. Several of my examples touch on issues of asymmetric power relations, which many people rightly care about. Dams generate power for some people but can also wreck the land, livelihoods, and culture of people who live around them. Likewise, it is hard to think about Aborigines or the Yanomami without also thinking about the dark side of settler and extractive capital. Even the word *Anthropocene* is politically charged. Is it "man," "humanity," that is to blame for global warming, species extinctions, and so on, or is it the West, the Global North, capital? Hence other names, like Capitalocene or Plantationocene.[2] I have mentioned these concerns in passing, but why have I not dwelled on them?

It is not that I don't care about them. But, as above, (a) I have nothing original to add on conventional political topics. There are shelves full of books on inequality already; the world does not need me to add another. And, again, (b) I am after something else. The politics of this book is a politics of practices, not social relations.[3] I have been trying to sketch out a different and unfamiliar pattern of poetic practice in the world — drawing sometimes on the practices of the marginal and the oppressed — to illuminate contrasts with ways of going on, enframing, that we already know too well.

Having said that, of course, it is entirely possible to fold our story back onto more conventional political considerations. In the realm of social relations, acting-on points, as usual, to inequity, hierarchy, and domination (as mentioned at the end of chapter 7, on fire), while acting-with points to radical democracy. I have discussed that elsewhere from a cybernetic perspective, including nonhierarchical solutions to the practical problems inherent in a symmetric politics (Pickering 2014). We can complete the loop here by noting that poetic approaches to wicked problems typically entail the evolving and more or less symmetric participation of all the interested parties (Pickering, forthcoming b).

. . . .

The academy has not been auspicious to relationality . . . the academy, taken as a whole, has been part of the occupying ontology. —ARTURO ESCOBAR, "Sustaining the Pluriverse"

The focus of this book has been on worldly practices like farming and civil engineering, but, in closing, one way to crystallize what is at issue is to think about education.[4] How does what we have been discussing bear upon educating our children and future generations? My own education and that of my children has amounted to an indoctrination into enframing. Schools teach about a knowable, predictable, and calculable world and how that knowledge can be put into practice. The current emphasis on so-called STEM subjects — science, technology, engineering, and mathematics — just intensifies this dualist enframing bent. Conversely, a good education (in, say, Britain, the United States, or Europe) from the earliest days at school up to a first degree at university and very probably a PhD as well, would not touch on anything poetic. Mine certainly did not, nor did my general experience of working in some of the world's best universities since then. It was only my idiosyncratic research interests that led me down the paths explored here. In terms of my yin-yang metaphor, modern education is tilted very strongly toward the yang pole and away from yin.

What then is to be done? Evidently, if we want to equip future generations with a lively sense of poiesis and the possibility of acting with nature, we need to rebalance the curriculum, and, in principle at least, that could be straightforward: we just need to teach students about the sorts of topics discussed here and in *The Cybernetic Brain* (Pickering 2010). At secondary-school level, for example, courses in mathematics could include discussions of complexity and cybernetics as leading into an ontology of emergence. Science courses could include a component on relations between science, enframing, and the dark side of the Anthropocene. Students could themselves experiment with natural farming on a small scale. Social science and history classes could explore the many reciprocal transformations of the human and nonhuman worlds. Classes in postdualist philosophy would help tie it all together. And as I said in *The Cybernetic Brain*, the arts can constitute a sort of nonverbal ontological pedagogy: interactive robots, as just one example, stage a nondual ontology of reciprocally coupled humans and nonhumans (for more on cybernetic art, see Pickering, forthcoming a).

Much the same could be said of university-level education, but thinking about this also brings the problem of disciplinary specialization into focus. One could teach adaptive management in engineering courses, say, without making any connection at all to more-than-human philosophy, natural farming, Aboriginal fire techniques, animism, cellular automata, psychiatry, and cybernetic management. The structure of both undergraduate and postgraduate education inhibits the sorts of cross-disciplinary paradigmatic interconnections I have tried to make visible here. Any poetic threads are liable to be lost in a dualist sea.

To counteract this fragmentation would require institutional as well as substantive change in the curriculum. One way to mark and reinforce inter- and antidisciplinary cross-connections would be the creation of academic centers and degree programs to bring together poetic research in a wide variety of fields and to organize a coherent curriculum spanning the topics just mentioned and more.[5] This would be a way to put poiesis as a paradigm — an alternative paradigm to the usual dualist disciplines — on the pedagogical and scholarly map. I cannot think of a good name for such a grouping — poetic studies? emergency studies? (Pickering 1995a) — but it would certainly be a way to foster the growth of a citizenry not oriented by default toward domination and ready instead to act with the world, not on it.

Another line of thought here moves from the content of education to its form. I have argued that poiesis offers us a constructive way forward in engaging with real-world wicked (exceedingly complex) systems, so a key step forward in poetic education might be to organize teaching around specific wicked and open-ended problems and strategies to tackle them. Against a background of the sort of poetic curriculum just outlined, such courses could serve as a first-hand introduction to performative dances of agency as a constructive approach to wicked problems, and perhaps even to the choreography of agency as a novel class of solutions.

This is not the place to go into detail, but the contrast we need is with conventional pedagogic approaches that focus on the transmission of established knowledge, examined by the rehearsal of the same and, in STEM subjects, solving assigned problems with solutions already known by the teacher. The poetic alternative would instead center on student experiences in tackling problems that have no such unique, pre-defined or foreseeable solution. In the social sciences and humanities this is already part of standard practice. In my own teaching, I often invite students to write creative essays on novel topics of their own choosing, somehow related to themes of the class. The added ingredient might be simply to encourage students to understand their own exploratory practice in the light of the sort of poetic curriculum just discussed.

In science subjects this sort of approach is gaining momentum in the form of classes centering on the design and actual making of new artworks, artifacts, and structures with an orientation to the future and to social context. These can serve both as opportunities for students to put technical knowledge to work and to alert them to the underappreciated performative and experimental — poetic — aspect of engaging with complex systems in the real world, again on the lines opened up elsewhere in a poetic curriculum. A new angle on STEM education.[6]

In the end, then, it is thus eminently possible to conceive a pedagogical rejigging of both the substance and form of education to foreground poiesis rather than enframing — another gestalt switch to add to the others this book has been about, and another key move toward acting with the world.

Notes

Epigraph 1. Horkheimer and Adorno, *Dialectic of Enlightenment*, 4.

Epigraph 2. White, "The Historical Roots of Our Ecological Crisis," 155.

Epigraph 3. Maitlis and Lévy, *Newsnight*, BBC2.

Epigraph 4. Blair and Hague, *A New National Purpose: Innovation Can Power the Future of Britain*, 4.

Epigraph 5. Feyerabend, *Science in a Free Society*, 73.

Epigraph 6. Shiva, *Staying Alive*, 54.

Epigraph 7. Horkheimer and Adorno, *Dialectic of Enlightenment*, 3.

1   Here is a quick list of some leading ways of thinking our way out of dualism. Leveling-up: emphasizing that matter is more interesting and lively than dualism credits — for example, object-oriented philosophy and new materialism — on the latter, Coole and Frost (2010), Jane Bennett (2010) on "vibrant matter" and distributed agency. From a different angle, monism: philosophically, one can assert that humans are built from just the same stuff as the rest of the world, and that our apparently exceptional properties (like consciousness) are themselves emergent from matter. Or there is the "extended mind" thesis — the idea that our cognitive processes themselves run through and depend on noncognitive and nonhuman materials as "scaffolding" (Clark 2001; Hutchins 1995). Scott Gilbert (2023) offers a fascinating biological account of the sympoiesis of organisms and environments. The Copenhagen interpretation of quantum mechanics points to a coupling of subjects and objects that offers an opening for a more general nondualist ontology (Barad 2007). My approach here is closest to work by Donna Haraway (2003, 2004, 2008) and Anna Tsing (2012, 2015) in science and technology studies on human-nonhuman couplings and to the actor-network approach of Bruno Latour (1987, 1993), Michel Callon, John Law, and others. Adrian Franklin (2023a) collects a range of key essays from a generally STS

perspective. Another tactic is to invoke the sort of nondualist and Indigenous knowledges (Escobar 2017; Law 2015; de la Cadena and Blaser 2018; Viveiros de Castro 2012) discussed in various guises as this book goes on, including Daoism and animism.

2   Note that this sense of agency — as centered on action, shared with nonhumans — differs from a more usual exceptionalist sense which identifies agency with putatively unique characteristics of the human race like will and intention. For an extended discussion of the concept of agency as used in this book, see Pickering (2023). See also Plumwood (2001).

3   Authors who touch on alternative patterns of action along the lines suggested here would include Heidegger ([1954] 1977), the contributors to Pickering (2008), Lorimer (2017), Scott (1998, 2017, 2020), Haudricourt (1969), Cuntz (2014), Haraway (2003), Tsing (2012, 2015), Hinchliffe (2022), Puig de la Bellacasa (2019), Keulartz (2012), Escobar (2017), and the cyberneticians discussed in Pickering (2010).

4   In *The Cybernetic Brain* (2010), I tend to speak of "revealing" rather than "poiesis."

5   How to build objects that resist responding to their contexts is an interesting question. The simplest answer is to use materials that turn out to be stronger and more enduring than the objects to be acted upon: concrete, rock, metal, glass, silicon. Of course, all these react eventually to their contexts, usually by degrading gradually or catastrophically in use. A more interesting answer involves using negative feedback to create a disconnection. My laptop contains a fan which switches on if its temperature rises too far, extending the range of indifference. I thank Pablo Jensen for pointing this out to me.

6   "Science" is a word with many meanings. In reference to "doing without science," I have in mind the sort of familiar modern sciences such as physics that are generally considered exemplary of science, and which contribute to engineering, say, in a straightforward calculative fashion. In Pickering (2010) I argued that cybernetics, and the sciences of complexity more generally, belongs to a different, nonmodern, paradigm.

7   The above remarks apply to what one might call finished science or the finished products of science — bodies of knowledge, instruments, machines. These are the elements of science that belong to the world of enframing, that make the world knowable, calculable, and predictable. But if we look upstream from these products to scientific research or what Bruno Latour (1987) called science-in-the-making, a different picture emerges. Genuine scientific research more or less by definition, is poiesis: open-ended performative finding out — try it and see. As mentioned, I actually first arrived at the concept of a dance of agency in examining scientific research practice (Pickering 1993, 1995a). On the one hand, then, we have scientific research as poiesis; on the other, finished science as the handmaiden of enframing. The way to reconcile these is simply to recognize that the overall *aim*, the guiding telos, of scientific research, is to make the world dual — to produce an instrument that can perform indifferently to its surroundings in the mode of enframing, or a piece of dualistic knowledge that can

stand independently of its creator. In general, then, poiesis as research practice can, in fact, aim at and sometimes arrive at enframing as its technical product. The point to remember in the following chapters is therefore that in discussions of science and enframing we are concerned with finished science, not science-in-the-making.

8   Heidegger's word is "frenziedness." He speaks of "the frenziedness of ordering" and "the frenziedness of technology" ([1954] 1977, 236, 237). "Frenzy" is also Kopenawa's (Kopenawa and Albert 2013) word to characterize Westerners exploiting the Amazon (see chapter 8).

9   "I was shocked when . . . I saw the pictures of the earth taken from the moon. We do not need atomic bombs at all [to uproot us] — the uprooting of man is already here. All our relations have become merely technical ones" (Heidegger 1981, 59). The contrast with the many expressions of techno-optimism and planetary consciousness that usually go with pictures of the earth from space is striking.

10 Elsewhere, I discuss the way in which mainstream developments in information technology and artificial intelligence explicitly aim to disengage users from the world, with self-driving cars as an obvious example (Pickering 2019). Conversely, I also discuss the ways in which cybernetic technologies have aimed at engagement. On engagement and disengagement in our relations with animals, see Pickering (2021).

11 When I first wrote these sentences (August 18, 2020) much of Beirut had just been destroyed by the accidental explosion of an enormous quantity of ammonium nitrate (a fertilizer as well as an explosive).

12 Wicked systems are much the same as the "exceedingly complex systems" that Stafford Beer (1959, 18) defined as the subject matter of cybernetics. On wicked systems and the cybernetic method, see Pickering (forthcoming b).

13 For a sustained account of the technological value of traditional ecological knowledge, see Watson (2019).

14 I do not want to overemphasize this theme of going-back. It does not apply, for example, to my interest here in contemporary sciences of cybernetics and complexity.

15 Elsewhere, I explore the issue of ontological hybridity in more depth in the case of cybernetics (Pickering 2013a).

1. EELS: THE DANCE OF AGENCY

An earlier version of this chapter appeared as Pickering (2005a).

Epigraph 1. Acts of John, The Gnostic Society Library: 26.

1   In a fascinating discussion of ontological politics, C. Jensen (forthcoming) extends the themes of Pickering (2005a) to bring eel history up to date. It has to be said that his eels are not my eels: his focus is on European eels in the sea, not Asian ones in American waterways.

2   On our side of the dance, all the strategies have focused on cutting ourselves off from the world: variations on quarantine, handwashing, masks, "social distancing," vaccines.

2. THE MISSISSIPPI: ENFRAMING AND LETTING GO

An earlier version of this chapter appeared as Pickering (2008).

Epigraph 1. General Thomas Sands, quoted in McPhee, *The Control of Nature*, 20.

Epigraph 2. Holling, *Adaptive Environmental Assessment and Management*, 134.

Epigraph 3. Schivelbush, *The Railway Journey*, 131.

Epigraph 4. Ghosh, *The Nutmeg's Curse*, 144.

Epigraph 5. Huxley, *The Doors of Perception*, 37.

Epigraph 6. David Muth, quoted in Schwartz, "A Mini-Mississippi River May Help Save Louisiana's Vanishing Coast," 6.

1   I thank Dawn Coppin for bringing this book to my attention.
2   On Native American nomadism, see Cronon (1983), though his focus is on New England.
3   "Falling height" is evidently a translation into English of the German for "potential energy." Kinetic energy actually would be more appropriate to train crashes.
4   Many levees failed in the hurricane, the effects of which were intensified by another of the ACE's works, the Mississippi River–Gulf Outlet (MR-GO) canal.
5   The only exception to this that I have found is an op-ed by a geophysicist and disaster expert in the *Washington Post* mentioned by Elizabeth Kolbert (2019, 22), who said that "politically, it was a non-starter."
6   Thanks to Lisa Asplen for this phrase.
7   The ACE continues with a typical enframing approach. "In 2013, the US Army Corps of Engineers spent $8.4 million on a dune along the beach, to shield the reservation against the fast encroaching tide." Already a Mississippi style dance of agency as arms race is emerging:

> Shortly after [2013] three storms damaged the dune. The Corps repaired it in 2018; storms wrecked it again. The Corps is now working on an even bigger stretch of dune — 25 feet high, 200 feet wide at the top and 4,000 feet long, protected by a 75-foot-wide stone revetment to hold the sand in place. Up close, the structure, which will cost as much as $40 million, looks less like a dune and more like a rampart. . . . Despite its massive proportions, the Corps predicts the dune will need rebuilding within ten years. "Nature's going to take over at some point," Mr. Ayala [an ACE engineer] said. (Flavelle 2022a)

8   The disappearance of the land behind the levees is another dark side of enframing, drawn out in time, rather than sudden as in the case of flooding.

9   "Brad Barth . . . the program manager for the project said that diversions are nothing short of revolutionary for those who manage the mighty waterway. For more than 100 years, flood protection was the priority, placed far above any consideration of the health of the wetlands" (Schwartz 2020, 3).

10 For much more on this project and a much less sanguine assessment of its threat to local livelihoods, see Coastal Protection and Restoration Authority (2020); and Rich (2020).

11 Ironically, the funding for this poetic diversion of the river comes from BP's settlement for the damages caused by the 2010 Deepwater Horizon oil spill — a massive enframing project with a truly catastrophic dark side (Schwartz 2020, 3).

12 Ben Fatherree (2004) outlines the history of numerical simulations, which are not necessarily superior to analogs. For example, a physical model of a Mississippi diversion just mentioned shows turbulent flows unexpected from numerical simulations (Schwartz 2020, 4). "Physical models continue to be irreplaceable, especially where turbulence and/or severe waves, factors poorly understood numerically, are important considerations. In the 1990s, in fact, use of physical models at WES actually increased" (Fatherree 2004, 199).

13 See the preceding note.

3. EROSION: POIESIS

1   Erosion control here involves fixed structures like dams that dominate waterflows, like the levees in the previous chapter, and in that sense, this example has a hybrid form, combining enframing technology with a poetic stance. I discussed this sort of hybridity in the introduction and will return to it in the final chapter. But I refer to erosion control as poetic because, as stated, it directs attention toward rather than away from the dance of agency. For more on this sort of hybridity and gestalts, see Pickering (2013b).

2   I make this distinction here because it turns out that when I talk about poiesis, people tend to equate it with skill.

4. THE COLORADO

Epigraph 1. Holling, *Adaptive Environmental Assessment and Management*, 2–3, 136, 14.

1   On the construction of the Glen Canyon Dam discussed in this chapter, see Rusho (2008, 527): " to the . . . Project Construction Engineer . . . the principal dilemma confronting him in 1956 had nothing to do with [designing and building] the dam but rather with the questionable location for the construction town."

2   Of course, as mentioned in the introduction, enframing has a fractal structure: all sorts of contingencies precipitate all sorts of small poetic dances of agency

in the real-time of construction, but these are subordinate to the overall design.

3   For a catalog of disastrous dam failures, see Perrow (1984, chapter 7). Many people would think here of the Vajont Dam catastrophe in Italy in which nearly two thousand people were killed in 1963, though that was caused by a landslide and not the collapse of the dam itself. The tide of poisonous red sludge that escaped from an alumina reservoir in Hungary in 2010 sticks in my mind for some reason (McCarthy 2010). In January 2019, the failure of the Brumadinho Dam in Brazil led to around 270 deaths. For the disastrous collapse of two dams in Michigan in May 2020, see Gray and Bosman (2020). On a glacier collapse which washed away a couple of dams and killed many in India, see Mashal and Kumar (2021).

Not all dam disasters are terminal. In the first artificial flood on the Colorado River discussed below, some of the excess water was discharged through purpose-built spillways — which immediately began to degrade dangerously and have never been put to use again. For a movie about this, see Aronson (1983). The subtitle of the movie is *Glen Canyon Damn Nearly Busts!* For a recent example of another spillway near-disaster in Oroville, California, see Nagourney and Fountain (2017); and Cox (2023). The population of Whaley Bridge in England was evacuated in August 2019 when overflowing water degraded a dam spillway: BBC News (2019). A July 26, 2020, newspaper article, on the history of the Kariba Dam on the Zambia/Zimbabwe border, states, "Over six decades of the waters' rushing through it, tumbling over it and crashing down the other side have eroded the dam's foundations and carved a pit at its base. Its plunge pool is now a 266-foot-deep crater" (Serpell 2020, 2–3). The Kariba Dam reappears at the end of chapter 5.

4   The classic example of performative excess would be carbon dioxide production and global warming as the unintended by-product of power generation.

5   A widely recognized form of such degradation by dams is the blocking of salmon runs. Salmon "ladders" are a poetic fix, in that they rely on the agency of the fish rather than thwarting it. For a review of possible environmental disturbances by dams and strategies of dam operation to address them, see Richter and Thomas (2007).

6   Pickering (forthcoming b). Similarly, elsewhere in ecosystem restoration, "the Everglades management confronts extremely complicated problems, many of which are 'wicked' . . . with singular responses" (Gunderson and Light 2006, 329).

7   Adaptive management is a general strategy that couples "learning to manage" and "managing to learn," foregrounding dances of agency in an obvious way. According to Lisa Asplen, the adaptive management approach to ecosystems was first laid out in Holling (1978), which presents a range of foundational studies. A later influential text is Lee (1991), based on a study of trade-offs between salmon and electricity in the Columbia River basin in the United States. Julian D. Olden et al. (2014, 176) note that "Large-scale flow events [central to the AMP, below] have entered the mainstream of water-resource management over the past decade and

the public profile of scientists and managers seeking ways to promote ecological sustainability using systematic adaptive management . . . is much higher than ever before." I thank Lisa Asplen for introducing me to the adaptive management of the Colorado and making the connection to the dance of agency, also for later discussions and information. I am also grateful to Jim Rice for his very valuable comments on an earlier draft of this chapter, suggestions, and further information.

8   The AMP also included modifications to ramping rates for electricity production but I omit these details here.

9   The sediment now forms enormous "mud glaciers" in Glen Canyon, exposed by the falling levels of Lake Powell, sliding toward the dam and threatening power generation (Maharidge 2023). This is another example of the emergent performative excess of enframing.

10   "Interior announced in 2012 . . . the adoption of a flexible protocol through 2020 wherein each HFE is 'triggered' by significant inflow of sediment to the Colorado by major tributaries below the dam. . . . A flexible protocol timed to the observance of sediment inflow will probably lead to annual or even semiannual HFEs punctuated by 1 or more years where no controlled flooding occurs due to drought conditions" (Rice 2013, 422).

11   Thus the 2023 HFE took place when "unusually wet monsoon seasons in 2021 and 2022 left more sediment piled up in Marble Canyon near the confluence with the Paria than usual" (NASA Earth Observatory). Among different reasons given for the cancellation of HFEs between 2018 and 2023 was the fall in water levels in Lake Powell attributed to climate change (Maharidge 2023), as in 2021 (GCDAMP Wiki 2022).

12   Konrad, Warner, and Higgins (2012) survey experimental floods and steps toward adaptive management of five other US rivers.

13   Charis Cussins (1996) uses the term "choreography" to refer to the changing subject positions of women as they move through in vitro fertilization (IVF). Ludwig Fleck (1979) comments on the "symphonic" performance of the Wassermann test for syphilis by a multiplicity of people and reagents working together.

14   The stated "purpose and goal" of HFEs, as of 2018, was "to determine if sandbar building during HFEs exceeds sandbar erosion during periods between HFEs, such that sandbar size can be increased or maintained over several years" (US Department of the Interior 2018, 4). Adaptive management can be understood as "treating management strategies as *experiments* that are conducted to learn more about the ecosystem's processes and structures" (Asplen 2008, 169, quoting Bosch, Ross, and Benton 2003, 11). For more on these contrasting senses of "experiment," see Pickering (2016). In his discussion of adaptive management, Kai Lee (1991, 67) refers to it as "experimenting without a laboratory" and notes that it elicits effects that "cannot be stumbled upon in the laboratory" (68). On "wild experiments," see also Lorimer and Driessen (2014); Callon, Lascoumes, and Barthe (2009, chapter 3) discuss "research in the wild," but they are referring to something else — to nonscientists themselves doing science, conventionally

understood, away from the "secluded" realm of the laboratory. Many of Latour's classic studies focus on the detour through the lab (or some "center of calculation"): on Pasteur, see Latour (1983); on Lapérouse and Sakhalin Island, see Latour (1987, chapter 6); on river modeling (as in chapters 2 and 4), see Latour (1987, 230–32); on soil science, see Latour (1999). A political critique of Latour would be that he seems unable to imagine a way of going on that does not involve this enframing detour (Pickering 2009a).

15 Along similar lines, we can also think of dam *removal* as experimentation in the wild, finding out how downstream ecosystems will react to changes in waterflow, and I thank Jim Rice for pointing this out to me. On the removal of massive dams on the Elwha River in Washington State in 2012 and 2014, see Nijhuis (2014); and Castillo (2023). We can think of dam removal as another example of letting go (chapter 2) and rewilding (chapter 5).

16 On monitoring the effects of experimental floods, see also Konrad, Warner, and Higgins (2012). For an interesting discussion of monitoring and mapping at a planetary scale, see Lenton and Latour (2018).

17 Likewise, Konrad, Warner, and Higgins (2012, 782) mention the use of models in planning and evaluating experimental floods, but note that "prescribed releases can have unintended and unforeseen consequences that are as important to evaluate as hypothesised outcomes." See also the quote from Holling (1978) at the start of the next section.

18 In the same spirit, "models are indispensible because without them human misunderstanding persists, unaware of its errors" (Lee 1991, 62). For more studies of scientific modeling in other contexts, see P. Jensen (2021, 2022); and P. Jensen and Li Vigni (forthcoming).

19 In the light of these resonances and parallels, it puzzles me that I have yet to find any significant explicit recognition of adaptive management in the cybernetics literature and vice versa. Holling (1978, e.g., 144, 185) tends to speak instead of "systems" and "systems analysis." Kai Lee (1991, 144–47) discusses what he calls "cybernetic learning," but his sense of cybernetics is different from that at issue here.

20 Rather than the homeostat, negative feedback devices such as thermostats are often taken as paradigmatic of cybernetics. But such devices lack the exploratory aspect of the homeostat which I take to be characteristic of poiesis, and, as noted in the introduction, they typically serve to cancel out the agency of the environment in which they act rather than incorporating it.

21 Meretzky, Wegner, and Stevens (2000) discuss the Kanab ambersnail, an endangered species perceived to be threatened by the first HFE in 1996. Calculations suggested that snails could be washed away. The solution was to move significant numbers of them higher up the bank immediately before the flood. In the end, no damage was done to the snail population.

22 Like drilling for oil beneath the Gulf of Mexico (Deepwater Horizon) — there is no happy medium between drilling and not drilling. "Regional developments or

policies are still designed within an economic context and reviewed only after the fact for their environmental consequences" (Holling 1978, 6).

23 Konrad, Warner, and Higgins (2012, 783) remark that "prescribed releases may not function as would be expected.... In this case environmental flow prescriptions ... can be implemented incrementally with 'stopping rules' in case of negative outcomes." On potential damage as emergent, Theodor Melis et al. (2011, 3) note that since the 2008 HFE, "The large increases of rainbow trout documented in the Colorado River ... may adversely affect adult populations of endangered humpback chub." The tentative quality of adaptive management made it possible to suggest a remedy: "Managers ... might choose to alter the timing of future HFEs to try to reduce the rainbow trout response" (4).

24 The idea of moving the sediment by pipeline has recently been resurrected in response to the mud glaciers mentioned in note 9 of this chapter.

25 A comparable slurry pipeline has in fact been built as part of efforts to restore lost land in Louisiana. The poetic counterpart is the construction of diversions on the Mississippi discussed in chapter 2 (Kolbert 2019).

26 Berkes, Colding, and Folke (2000) make many productive connections between examples of "Traditional Ecological Knowledge," complex systems, and adaptive management: "Adaptive management, like many traditional knowledge systems, assumes that nature cannot be controlled and yields cannot be predicted. Uncertainty and unpredictability are characteristics of all ecosystems.... In this sense, adaptive management can be seen as a rediscovery of traditional systems of knowledge and management" (1260).

27 In the list of ancient poetic technologies, it is appropriate to mention here the Dujiangyan Dam in China and the Grand Anicut in India, both two thousand years old. Both are key elements of irrigation systems, and their common feature is that they act differently according to the amount of water flowing past them (and thus diverting floods away from the irrigated area). These structures, too, act with rather than on nature. Dujiangyan reappears in the discussion of Daoism at the end of chapter 6.

5. WATER: A NEW PARADIGM

Epigraph 1. Frans Vera, quoted in Kolbert, "Recall of the Wild," 4.

Epigraph 2. Book of Genesis, 1:26, 27.

Epigraph 3. White, "The Historical Roots of Our Ecological Crisis," 1206.

1 Here I follow Zwart's (2003) periodization and discussion.

2 Zwart (2003, 109) cites Pliny the Elder as describing the watery landscape as bleak and appalling, but Scott (2017) argues against the reliability of such judgments.

3 On science in relation to the behavior of water itself: "Around the turn of the [twentieth] century ... Lorentz ... was asked to predict the tidal effects caused

by the closure of the Zuyderzee. . . . The extensive calculations this involved would take him eight years. From the engineer's point of view, the project was an astonishing success" (Zwart, 2003, 120). Wiebe Bijker (2002, 577–78) mentions Lorentz's work in the 1920s and also physical models of the Zuider Zee (his spelling) that parallel the models of the Mississippi discussed in chapter 2. He briefly discusses the utility of experience with a model in closing a breach left by the 1953 flood (discussed in this chapter), as well as an example of the failure of such modeling (Bijker 2002, 578n19). As Bijker mentions, Bruno Latour (1987, 230–32) gives a nice sketch of Dutch river modeling at the Delft Hydraulics Lab. The Professor Bijker that Latour mentions is Wiebe Bijker's father. Latour gives no citation, so I assume he got this information from Wiebe. Latour is very clear that through the model "Professor Bijker and his colleagues *dominate* the problem, *master* it" (231, original emphasis).

4 One way to limit the size of floodplains and thus increase flood risk is building dikes. Another is canalizing rivers. The latter was a problem in the disastrous floods in western Europe in July 2021. "Where the Erft passes Erfstadt," one of the towns in Germany where flooding was most devastating, "it is no longer a naturally flowing river but more like an artificially straightened canal. It flows much faster here than elsewhere and lacks the natural floodplains that could deal with overflow" (geographer Matthias Habel, quoted in Tenz and Oltermann 2021). The absence of any comparable flooding in the Netherlands in the same period is credited to the Room for the River project (Erdbrink 2021).

5 Also, "a pair of rare white-tailed eagles [nested] below sea level, displaying behaviour unanticipated by ornithologists," and "cattle and horses at OVP display demographic structures, herd dynamics and individual coping mechanisms that confound experts on their domestic kin" (Lorimer and Driessen 2014, 175).

6 We can complicate the story of science and the OVP. Frans Vera, the leading figure in the rewilding of the OVP, often portrays it as pointing to a shift of "paradigms." The traditional paradigm in conservation to which he refers assumes that the natural state of the European landscape is dense forest inhabited by relatively few animals. Vera advocates a novel paradigm in which birds and herbivores are understood to maintain a relatively open landscape and argues that developments at the OVP support the latter. The introduction of horses, cattle, and deer fit in with Vera's agenda. In that sense, the OVP is a rather conventional scientific experiment, and the validity of Vera's argument is disputed on various conventional grounds (Fisher 2019). Most accounts, however, including some of Vera's, emphasize instead the OVP as open-ended experimentation in the wild aimed at finding out how the OVP will evolve. Lorimer and Driessen (2014, 174) quote Rewilding Europe (2012, 2) as asserting that "rewilding as a concept . . . does not aim at the fixed conservation of particular species, habitats or a priori lost landscapes, but rather opens for the continuous and spontaneous creation of habitats and spaces for species." Lorimer and Driessen (175–78) continue:

What is perhaps most surprising and different about OVP is the lack of prediction and management that has taken place. Until recently there have been no targets, no models and no explicit action plan.... OVP became famous as a source of surprises and those interested in its ecology were keen to nurture and learn from its inadvertent ecological processes.... He [Vera] argues that this has generated a range of surprising ecological events and new ecological knowledge that challenges the existing paradigm.... Letting go of a safely composed Nature might lead to economic risks and the local and global diminishing of biodiversity.... Although the contemporary ecology of OVP is presented as a test of Vera's hypothesis, in practice it is valued for its ability to surprise.

Likewise, Elizabeth Kolbert (2012, 5) states, "Michael Coughenour, a research scientist at the Natural Resource Ecology Laboratory at Colorado State University, was a member of ICMO.... 'I didn't see anything that looked bad to me,' he went on ... 'I think it's a great experiment to let it run and see what happens.'" We could think of two gestalts here, one of which picks out the scientific elements of rewilding, the other the openness to surprise and emergence. The latter is germane to our interests.

7 On relations between Heck cattle and Konik horses and their putative extinct ancestors, see Kolbert (2012), who explores the surprising origins of Heck cattle in Nazi Germany.

8 Erica Gies (2022) explores many examples of the acting-with, poetic paradigm, including diversions on the Mississippi (261), managed retreat (268), and "letting go" (mentioning Hurricane Katrina, 283). Most of her examples, however, focus on scientists as "water detectives" who know what water wants to do, in contrast to the experimental approaches discussed here. Marie Meilvang (2021) explores a recent shift in Denmark from treating rainwater in cities as a problem to be solved by drainage (enframing) to a resource for urban enhancement (poiesis), and includes a useful review of parallel changes around the world.

9 I thank Antonio Carvalho for encouraging me to think about this.

10 More specifically, Zwart (2003, 113–14) discusses the subsequent importance of Calvinism in legitimating and encouraging "ambitious and expensive reclamation projects." Zwart also argues that at a practical level, "only monastic orders were able to recruit a sufficient number of 'hands' in those days," and "monasteries played a prominent role in the medieval transformation of soggy wilderness into a place more pleasing to man and God. They played a large part in reclamation projects, just as they did in forest clearances elsewhere in Europe." Christianity, then, had the social organization to translate its beliefs into practice. Eventually one could speak of a "clash between a Christian, enlightened world, on the safe side of the dike, and an older world filled with pagan, Wotanesque reminiscences beyond it" (Zwart 2003, 111, 118). Apart from religious sources, of course, the secular technological developments in water management mentioned earlier were

central to Dutch water control. As another example, White (1967) argues that the mediaeval introduction of heavy ploughs was central to installing the stance of enframing in farming, as also discussed in the next chapter of this book.

11 Lynn White's argument has been foundational to subsequent discussions of ecology and religion and continues to be controversial (Whitney 2015). On White's account, the history of western technology and science is bound up with the Christian enframing of nature. The coming together of science and technology in the second half of the nineteenth century then drove the expansion of industry, which in turn elicited the "ecologic backlash [which] is mounting feverishly. . . . No creature other than man has ever managed to foul its nest in such short order" (White 1967, 1203–4). Chandra Mukerji (2002) explores the "material practices of domination" that characterized "Christian humanism" in the sixteenth and seventeenth centuries, especially the design of gardens in France.

12 The current pope has, of course, taken the name Francis, and has argued strongly that we should care about the environment and global warming (Francis and Pope 2015; Yardley and Goodstein 2015). On alternative Christianities, see also the quote from the Gnostic Acts of John at the start of chapter 1.

13 The Tonga were right socially, too. "Construction of the Kariba Dam . . . required the Tonga to move from their ancestral homes along the Zambezi River to infertile land downstream. Mr Scudder [an eminent dam consultant] has been tracking their disintegration ever since. Once cohesive and self-sufficient, the Tonga are troubled by intermittent hunger, rampant alcoholism and astronomical unemployment. Desperate for income, some have resorted to illegal drug cultivation, elephant poaching, pimping and prostitution. Villagers still lack electricity" (Leslie 2014).

14 On spirituality in the history of cybernetics, see Pickering (2009b, 2010, chapter 6).

6. NATURAL FARMING

Epigraph 1. Gregory Bateson, quoted in Harries-Jones, *A Recursive Vision: Ecological Understanding and Gregory Bateson*, 205.

Epigraph 2. Martin Heidegger, "The Question Concerning Technology," 224.

Epigraph 3, Bill Mollison, quoted in Scott London, "Permaculture: A Quiet Revolution," 2.

Epigraph 4. Masanobu Fukuoka, *The One-Straw Revolution*, 40.

Epigraph 5. Masanobu Fukuoka, *The One-Straw Revolution*, 19.

Epigraph 6. Masanobu Fukuoka, *Mother Earth News*, 4.

Epigraph 7. Martin Crawford, *Growing a Forest Garden*, 20.

Epigraph 8. Masanobu Fukuoka, *The One-Straw Revolution*, 125.

Epigraph 9. Masanobu Fukuoka, *The One Straw Revolution*, 145.

Epigraph 10. Gregory Bateson, *Steps to an Ecology of Mind*, 268.

Epigraph 11. Joseph Needham, *Science and Civilisation in China*, 85.

Epigraph 12. Masanobu Fukuoka, *The One-Straw Revolution*, 119.

1   Many of the following remarks apply equally to animal farming. For an important analysis of large-scale pig farming which brings out its enframing aspect nicely, see Coppin (2002, 2003, 2008). For a fascinating account of cutting-edge pig farming in contemporary Britain, see Hinchliffe and Ward (2014). These authors contrast two stances toward disease and immunity in pig farming. One focuses on maintaining closed disease-free systems (enframing as blanket exclusion); the other, on tuning the immunities of pigs entering new systems (poiesis as situated incorporation).

2   Thus, Siegfried Giedion (1948a, 130) treats the phrase "tiller of the soil" as synonymous with "farmer."

3   On the history of the mechanization of agriculture, see Giedion (1948b).

4   A progressive historical shift toward raising animals indoors in confined spaces — factory farming — has this same negative aspect of suppressing the agency of animals (Coppin 2008). A parallel recent shift to raising crops indoors is evident in aquaponics and "vertical farming" (Mageau. 2016; Severson 2021).

5   The history of this detour goes back to the work of Justus von Liebig in the mid-nineteenth century, but the importance of the chemical industry has increased markedly since WWII. GMOs depend heavily on laboratory science, of course. We can also think about experimental plantings and the statistical analysis of the effects of all sorts of variables. Giedion (1948b, 139) singles out the Earl of Leicester's estate ca. 1800 as the first experimental farm. For a recent example of the statistical analysis of different planting conditions, see Nafziger (2020). The moral of all this is that the transition to enframing in farming is not a simple on-off shift — we could think of layers, including the development of pure lines in plant breeding (Bonneuil 2016).

6   James Scott (2017) analyzes the rise in farming in terms of state formation rather than productivity. On problems associated with plowing (carbon dioxide release, erosion, lack of water retention) see Wreg (n.d.) and *Farming News* (2014).

7   Organic farming resembles conventional farming but abstains from the use of industrial chemicals. Permaculture aims at self-sustaining assemblages of perennials. On permaculture, see Mollison and Holmgren (1978) and Crawford (2010). On a broad approach to "regenerative agriculture" see National Resources Defense Council (2021).

8   This was brought to my attention by a very insightful essay by Thomas Smith (n.d.); I am very grateful to Smith for getting in touch with me. In 1987, *Mother Earth News* published an interview with the "Holy Trinity" of permaculture, who they list as Masanobu Fukuoka, Wes Jackson, and Bill Mollison.

9   Natural farming now has exponents around the world (*Ideas*, n.d.) especially in India (Tegta 2010; Rathore 2019; Aggarwal n.d.). For a review of natural farming in India, including a comparison with organic farming, see Rana and Manuja (2021).

10   The distinction between poiesis as a human stance in the world and as technique or technology can seem cumbersome, but thinking about this study first convinced me that it needs to be made. There are aspects of natural farming as technique that are not clearly apparent in Fukuoka's route to it.

11   In 1938, at the age of twenty-five, Fukuoka was hospitalized with acute pneumonia. Afterward he felt depressed and confused. One night, "I collapsed in exhaustion on a hill overlooking the harbor." When he awoke in the morning "a night heron appeared, gave a sharp cry, and flew away into the distance.... In an instant all my doubts ... vanished. Without my thinking about them, words came from my mouth: 'In this world there is nothing at all.'" (Fukuoka 1978, 8). A footnote to this explains, "To 'understand nothing' ... is to recognise the insufficiency of intellectual knowledge." We can connect this forward to Fukuoka's rejection of science. He himself paraphrased it as "Humankind lives in an unreal world separate from nature," and his subsequent development of natural farming grew from his desire "to demonstrate how my ideas could be of practical benefit to society.... My idea was to let nature have a free hand" (Fukuoka 2012, 4, 5). Enlightenment through exhaustion reminds me of nothing so much as Jack Kerouac and the Beats' version of Zen. On the other hand, Fukuoka's recollection that "in a single leaf, a single flower, I was moved to appreciate all the beautiful forms of this world.... When I viewed the world with an empty mind, I was able to perceive that the world before me was the true form of nature, and the only deity I would ever worship," (Fukuoka 2012, 5) reminds me of Aldous Huxley's (1954) description of experiencing the "suchness" of the world after taking mescaline for the first time, which he also related to Zen Buddhism. We will come back to Eastern spirituality.

12   Mastery in farming can require more rather than less human labor (Coppin 2002, 2003, 2008). Fukuoka's idea that we should do less was intended to apply to all spheres of life; he saw farming as a specific site to explore the possibilities.

13   "Fukuoka also implies that our fixation on control over nature has led us to assume visual order — the straight weeded rows of uniform fields — is superior farming" (Lappé 2009, x). This sort of visual order is precisely the target of James Scott (1998), as signaled in the title of his work *Seeing like a State*.

14   The pellets also help protect the seed from mice, voles, and so on; sowing while the preceding crop is in the ground also helps shield the seeds from sparrows (Fukuoka 1978, 51–52).

15   For a nice account of traditional rice farming in Japan, see Korn (1978, xxi–xxii). Like natural farming, traditional farming alternates between rice and rye or barley in each field over the course of a single year, so the striking differences are not in fact in this alternation. Overlapping the two crops, however, is part of what sets natural farming apart.

16   Permaculture is also a complex choreography of the agency of a large number of species growing together, but without the temporal punctuation of natural farming or the Colorado Adaptive Management Program (Mollison and Holmgren 1978; Crawford 2010).

17   This timing of operations in relation to worldly performances also appears in Scott's chapter on *metis* (1998, 311): European settlers in the New World "were told by Squanto . . . to plant corn when the oak leaves were the size of a squirrel's ear."

18   "The most commonly used chemical fertilizers . . . are used in large amounts, only fractions of which are absorbed by the plants in the field. The rest leaches into streams and rivers, eventually flowing into [Japan's] Inland Sea. These nitrogen compounds become food for algae and plankton which multiply in great numbers, causing the red tide to appear" (Fukuoka 1978, 82). We could think here of industrial farming as a dualist island of stability and the run-off of chemicals as an example of performative excess (Pickering 2017b).

19   On caring for and reanimating soil, see Puig de la Bellacasa (2019).

20   At a smaller scale, a letter from a farmer to my local parish magazine (Ayre 2020) comments on a similar lock-in: "EU legislation banned a chemical that was used on rape to control flea beetle . . . resulting in having to spray the crop more often with not so efficient chemicals. This means more input to the crop and less yield. The legislation has caused a reduction of about 70% nationally in the area of rape being planted, as it is no longer economical to grow." In his later years, Fukuoka (2012) became involved in efforts to reinvigorate deserts. Again this involved a lot of transformational work, scattering a multiplicity of seeds in clay pellets to see which would grow and begin a process of revegetation. Like the path to natural farming, one can see this transformation of the desert as, according to Fukuoka, a restoration of a prior state of the land. If the move from industrial to natural farming entails changing the land itself, we could note that the same goes for human subject-positions. As mentioned in note 11 of this chapter, only in extremis did Fukuoka discover his new poetic stance. The same goes for the decentered subject position that features in deep ecology (Naess 1973).

21   Fukuoka's writings on going back can be read in two ways. One is as recommending a "do nothing" approach in farming today. This, I think, is how most people who have adopted natural farming read him and it is the position I have indicated in the text. But his writings include passages where he clearly recommends going back tout court and adopting a very minimal lifestyle, living on the land without modern amenities like running water or electricity. This is how visiting disciples lived on his farm during his lifetime, and it is also, for example, consistent with his critique of education as a detour away from nature (Fukuoka 1978, 17).

22   Another way to think of this would be to think of the world as a unity but one which shatters into an endless sequence of pieces when one tries to grasp any single thread.

23 For a scientific account of pesticide use as an endless arms race, in many ways paralleling Fukuoka's arguments, see Hajek and Eilenberg (2018, chapter 1).

24 Fukuoka (1994, 28) tells an even more complicated version of the weevil-blight story, including, besides the elements just listed, clear-cutting, planting cedars, small birds, long-horned beetles, matsutake mushrooms, atmospheric pollution and acid rain.

25 C. S. Holling (1978, 26–27) reviews a study of the use of pesticides in Peru:

> The initial response to the insecticide treatment was a pronounced decline in pests and a 50 percent increase in cotton production. After two or three years, however, six new species of insects became as serious a problem as the original seven had been. The reason for the appearance of these new pests was the elimination of parasites and predators that were killed by the insecticides. Within six years the original seven insect pests began to develop resistance to the insecticide, and crop damage increased. In order to control this resurgence, the concentration of the insecticide had to be increased and the spraying interval reduced from two weeks to three days. As these control measures began to fail, the chlorinated hydrocarbons were replaced by organophosphates. But even with this change, the cotton yield plummeted to well below those realized before synthetic insecticides. The average yield in 1956 was the lowest in more than a decade, and the costs of control were the highest: the agricultural economy was close to bankruptcy.
>
> [In response] "This forced the development of a very sophisticated ecological control program that combined changed agricultural practices with the introduction and fostering of beneficial insects. Chemical control was minimized. These new practices allowed the re-establishment of the complexity of the food web, with the result that the number of species of pests was again reduced to a manageable level. Yields reached the highest level in the history of cotton production in the valley."

26 My reading of Holling (1973) is that Fukuoka's idea that pests and predators can be induced to come into a satisfactory form of equilibrium is not necessarily true. Of course, Fukuoka did not claim to have proved that; his claim was that he had found it possible in practice. I thank Malcolm Nicolson for discussion of this point.

27 In fact, the quotation is taken from one of Gregory Bateson's ([1959] 2000) essays on schizophrenia but it applies equally to his view of the environment.

28 On ecology and the environment, see G. Bateson (1968, 2000b), M. Bateson (2005), and Harries-Jones (1995).

29 We could note that this cascade of chemicals constitutes a kind of lock-in. Claire Brown (2021, 2) quotes a farmer concerned that "a season without dicamba [a weedkiller] would mean devastation for his farm." At the same time, weeds are evolving to resist commercial weedkillers including the most popular, Roundup (glyphosate), leading to "superweeds" resistant to weedkillers they have never been exposed to before (Brown 2021). The parallel to antibiotic-resistant microbes is clear, as is the involvement of science in both of these arms races.

30   André Haudricourt (1969) also links farming practices and ontological under-
      standings. He contrasts Western and Eastern approaches to farming as "direct,
      positive" and "indirect, negative" action (164) — in my terms, enframing versus
      poiesis (or *wu wei*, discussed in this chapter) — but associates the contrast with
      what is farmed. The iconic farm animal in the West for Haudricourt is the sheep,
      which requires constant attention from the shepherd, a stance of enframing. The
      iconic crop for him in the East is the yam, for which the farmer prepares the soil
      but then stays away to leave the plant to grow as it will, poiesis. Haudricourt's
      suggestion is that this contrast is fundamental and feeds into more general con-
      trasts between East and West, including philosophy: "Is it so absurd to ask if gods
      who ordain, the morals that regulate, the philosophy which transcends, have no
      connection with sheep, as traced through a predilection for slavery and capitalist
      methods of production? Or whether the morals which explain and the philoso-
      phies of immanence have not got something in common with yams, taro, and rice,
      in accordance with the production methods of ancient Asia and bureaucratic feu-
      dalism?" (172). Haudricourt puts rice farming in the same category as yams (165),
      which would make sense for Fukuoka's natural farming but hardly so for con-
      ventional rice-farming. He also makes an interesting parallel contrast between
      formal French and Italian gardens (think of Versailles) and naturalistic Chinese
      gardens and bonsai (165). For more on connections between gardening and West-
      ern modernity, see Mukerji (1994, 1997, 2002). I thank Michael Cuntz for bringing
      Haudricourt to my attention; see Cuntz (2014) for an insightful discussion.
31   I thank Brian Rappert, Giovanna Colombetti, Regenia Gagnier, and John Dupré
      for help in thinking how to formulate my thoughts here.
32   The quote continues: "'In pursuit of wisdom, one adds everyday. In pursuit of
      dao, one decreases everyday" (Laozi 2008, chapter 48). "This is meant to abol-
      ish intellectual learning, cognitive knowledge, and distinction-making reason-
      ing. Since the natural state is based on the awareness of the limits of conceptual
      knowledge, the more one pursues it, the farther he is from the *dao*. We need to
      abolish knowledge-based social guidance to restore the original spontaneous
      state" (Yu 2008, 8).
33   Along the same lines, Fukuoka also rejected the more positive elements of East-
      ern philosophy: "In the West natural science developed from discriminating
      knowledge, in the East the philosophy of yin-yang and the *I Ching* developed
      from the same source" (1978, 125).
34   Fukuoka (1978) disavowed affiliation with any organized religion: "I do not
      belong to any religious group. . . . I do not care much for making distinctions
      amongst Christianity, Buddhism, Shinto, and the other religions" (116). He did,
      however, write extensively on the interpretation of key Buddhist concepts (and
      interrelations with Christianity, etc.). His first book, self-published in 1947, was
      called *Mu: The God Revolution*. Various other self-published books also called *Mu*
      followed (Green, n.d.; Wikipedia entry on Masanobu Fukuoka). *Mu* (*wu* in Chi-
      nese, as in *wu wei*, discussed in this chapter) is a Zen Buddhist term of art for

"nothing" or "no-thing" — as in Fukuoka's description of his moment of Enlightenment (note 11 of this chapter): "I felt that I understood nothing" (1978, 8). From a different angle, "Among natural farming methods two kinds could be distinguished: broad, transcendent natural farming, and the natural farming of the relative world [the world as understood by the intellect]. If I were pressed to talk about it in Buddhist terms, the two could be called respectively as Mahayana and Hinayana natural farming" (Fukuoka 1978, 118; see also Fukuoka 1994, 91–93).

35   Jullien (1999, 11) states that *shi* "is a relatively common word" in Chinese, "generally given no philosophical significance." Nevertheless "it is not merely Chinese thought that might be illuminated — that is, the whole spectrum of Chinese thought which we know has focused on perceiving reality as a process of transformation. The logic of *shi* could even pass beyond peculiar cultural perspectives and thereby illuminate something that is usually difficult to capture in discourse: namely, the kind of potential that originates not in human initiatives but instead results from the very disposition of things. Instead of always imposing our own longing for meaning on reality, let us open ourselves to this immanent force and learn to seize it" (Jullien 1999, 13). Cuntz (2014, 119) also discusses *shi* in the present connection.

36   This mastery is exemplified in the often-repeated Daoist story of Cook Ting: "When I first began cutting up oxen, all I could see was the ox itself. After three years I no longer saw the whole ox. And now — now I go at it by spirit and don't look with my eyes. Perception and understanding have come to a stop and spirit moves where it wants" (Chuang Tzu 1964, 46–47). Krueger (2009) equates this sort of noncognitive mastery with skill, but I argued in chapter 3 that skill is a human-centered concept, whereas my aim here is to draw attention to decentered performative relations between people and things.

37   In our first example (chapter 3), the forest engineer explores the *shi* of water, vegetation, soil, and so on, discovering how they will perform when configured this way or that. The Glen Canyon Dam story has just the same form, experimenting with waterflows, hoping to find a propensity to restore downstream sandbanks and ecologies. Room for the River allows the Rhine to display its propensity to act, just as the rewilding of Oostvaardersplassen allows nature to do so (chapter 5).

38   In his discussion of the origins of Buddhism, Alan Watts (1957, 57, 58, 61) says that "complete action is ultimately free, uncontrived, or spontaneous action, in exactly the same sense as the Taoist *wu-wei*," and he makes a contrast with patterns of action that generate karma, of which he says, "Man is involved in *karma* when he interferes with the world in such a way that he is compelled to go on interfering, when the solution of a problem creates still more problems to be solved, when the control of one thing creates the need to control several others. *Karma* is thus the fate of everyone who 'tries to be God,' He lays a trap for the world in which he himself gets caught." Karma, then, is the concomitant of enframing, as in the Army Corps of Engineers' endless attempts to control the Mississippi River. Interestingly, Watts remarks that "this is really a simple problem

of what we now call cybernetics." He has in mind here a control mechanism like a thermostat which hunts back and forth, never quite coming to rest. In the mid-1950s, Watts collaborated with Gregory Bateson in his well-known work on the double-bind and schizophrenia (Pickering 2010, chapter 5).

39 She does not go into any detail, but among several interesting examples, Sandra Wawrytko (2005) mentions the first artificial flood on the Colorado River and the Dujiangyan Dam, discussed later in this chapter, as instances of *wu wei* in action, and levees on the Mississippi as *wei*: "doing" as enframing. "To undo the damage of dualism, the Daoist Sage neither acts upon (manipulates/exploits), nor is acted upon (is manipulated/exploited) by the environment, but rather *interacts* with it, doing the dance of Dao" (Wawrytko 2005, 95).

40 I thank Professor Shihua Li of the Philosophy Department of Nanjing University for first telling me about the Dujiangyan Dam and making the connection to my discussion of dances of agency and the Mississippi River. See Li (2012). Shanghong Zhang et al. (2013) offer a useful and relatively comprehensive technical description of Dujiangyan's components and functioning; I thank James Hevia for drawing this paper to my attention. Joseph Needham (1971, 288–96) is another valuable and simpler description (under the headings of Kuanhsien and Tu-Chiang Yen). For a parallel example, on another thousand-year-old dam, the Grand Anicut on Kaveri River in Tamil Nadu, India, see Bijker (2007, 111–14).

7. A CHOREOGRAPHY OF FIRE

Epigraph 1. Bill Gammage, *The Biggest Estate on Earth*, 141.

Epigraph 2. Bill Gammage, *The Biggest Estate on Earth*, 142.

1 Many of the thatched cottages in my English village have burned down over the centuries, often being reconstructed with tiled roofs — including my house. "The European landscape ideal is a Garden, with everything (and everyone) in its proper place. Fires broke out during times of unrest. . . . Eliminating fire belonged with the abolition of plagues and banditry" (Pyne 2020, 96–97). "As Europeans took control of [Australia] they banned burning. . . . 'The European mind-set was to be totally scared of fire'" (Fuller 2020, quoting Jeremy Russell-Smith). Australian, like European, elites, "found in the contagion of fire a practice to condemn, an enemy to fight, and a symbol of reckless and extravagant waste" (Pyne 2020, 49).

2 Speaking of the early 2000s: "The truly malicious irony, however, was reserved for the most fire-protected bush, the public domain. Here decades of attempted fire exclusion in the name of good government, right thinking, and the protection of timber and catchments had let nature stockpile combustibles like a bower bird seeking to entice flame. . . . Like levees thrown up to protect a food-plain, control efforts had eliminated small, nuisance events, only to worsen the setting for the massive ones. Instead of routine fires washing away the encrust-

ing fine fuels, catastrophic fires overflowed and drowned whole forests" (Pyne 2020, 85).

3   "Experts agree that prescribed burns ... are a key to reducing the severity of wildfires [in the US] in the future. ... But experts also stress that there needs to be more federal and state legislation that prioritizes this technique. ... Another important step is ... to remove dead trees and other fuel" (Schagrin 2021, 5).

4   "The absence of periodic 'cleansing' blazes ... often increases the size and severity of fires beyond managerial control" (Asplen 2008, 168). Aboriginal-style "defensive burning" is now officially encouraged in Australia (Fuller 2020).

5   Jill Cowan (2020) reports on a new interest in Native American uses of fire in the United States, observing that "until recently, it has been mostly dismissed as unscientific." Marcia Macedo and Valéria Pereira (2020) discuss Indigenous fire management by the Kuikuro in the Amazon but mention only defensive measures, complex earthworks such as ditches, roads, and dikes, plus fire-brigades. For an insightful account of a multi-functional "cultural burn" conducted by Native American tribes in a sequoia grove in California, exemplifying themes discussed below, see Robbins (2024).

6   Gammage's book has entered into and fed a swirl of arguments and controversies about Aborigines and fire, concerning topics as various as the extent of Aboriginal fire, Gammage's white-centric approach and his account of the Dreaming, and highly charged and political debates about burning in Australia today (e.g., Coulice and Murphy 2015; Fantin 2013; Neal 2012; Wynter 2015). For an earlier move in the controversy on Aboriginal fire, see Horton (1982). I am concerned with the form of Aboriginal fire practices. I think no one questions Gammage on this topic, and I follow him as offering the most detailed range of examples one could hope for. I could just note that the "scientific" position in the controversy is that if features of the landscape can be understood as independent of human causes they should be. Conversely, the "historical" position (e.g., Gammage) is that if features can be explained as the result of human action they should be. There is a parallel here with arguments about rewilding (chapter 5) and whether the ground state of the European landscape is closed-canopy forest supporting few large animals, or open-canopy maintained by a relatively large number of large animals. In each case, it seems to be possible to amass evidence and assumptions to support either side, in a way which is familiar to me from my studies of the discovery of the weak neutral current in particle physics (Pickering 1984b). There, suitable configurations of assumptions and data-processing patterns pointed convincingly either to the existence or the nonexistence of the phenomenon in question.

7   "When in summer fires were lit varied with how damp the country was, how big people wanted patches to be, which food plants were ready, which seed had germinated, whether game was to be attracted or moved, and so on. Grass-seed eaters burnt after harvest, usually in late summer, relying on timing and damp plant

bases to fetter the flames; tuber eaters mostly in early summer once tubers matured. Wet country was lit whenever fuel was dry enough, which meant mostly in summer." For more examples, see Gammage 2011, 56, 163, 164, 173, etc.

8 On similar linkages: "Songlines also recite countless ecological signals to people and animals: when coral trees flower it is time to dig crabs, when a bird sings a grub lays eggs, when the west wind blows blue-tongue lizards emerge and women dig for honey ants, when march flies appear crocodiles lay eggs, when the blackwoods flower northwest Tasmanians hunted muttonbirds. Such specific associations are necessarily local and seasonal, varying each year" (Gammage 2011, 135–36); compare to Squanto in chapter 6, note 17, of this book.

9 Similarly,

> Scott makes clear how crucial timing was. Whereas fuel control was a fire plan's baseline, timing was its end point, considering both fuel and purpose. In north Queensland early Dry fires promote some shrubs and grasses, late Dry fires promote others and herbs. When to burn grass might hinge on its varying growth from year to year, or on associated tubers or annuals, some killed by fire, others needing it to flower, seed or compete. These needs are easily unbalanced. For example badly timed fire promotes unpalatable perennials like Blady Grass or Bunch Spear Grass at the expense of fodder grasses, and has let Spinifex expand its range to become "the most common organism in the dry three-fifths of Australia." Yet plant communities embracing different fire responses thrived in 1788. Multiply this by Australia's 25,000 species, and a management regime of breathtaking complexity emerges. (Gammage 2011, 167)

10 "In a heath near Kiama (NSW), ground parrots needed fire every three to seven years to balance food and shelter. In 1788 they got this, but after 1788 they got infrequent hot fires, and by 1968, they had died out" (Gammage 2011, 17). From the opposite angle, fire controlled populations of insects by "destroying a vast number of insect eggs and larvae" (Gammage 2011, 182).

11 "Mitchell recognised that grass to lure game was 'worked from infancy.' West of the Macquarie Sturt found 'the grass, which had been burnt down, was then springing up most beautifully green, and was relished exceedingly by [our] animals.' Kangaroos, George Bennett advised in 1832, 'like the cattle, frequent those places where the grass, having been recently burnt, they meet with the sweet young herbage'" (Gammage 2011, 177).

12 "Skilful burning has kept the forest dense, the grass open, the game convenient. If people spread enough templates around, they could always hunt somewhere" (Gammage 2011, 92).

13 See also Pyne (2020): "In the Alpine fires, country people often objected to what they saw as ignorant outsiders who blundered around the landscape unaware of the peculiarities of the scene, substituting mechanical muscle for hard-won local knowledge" (137). "This failure comes precisely because those institutions

do not, or cannot, function at a landscape-scale sufficiently fine-grained with the opportunism and nose-to-the-ground sensitivity to match the variety of the world under their care. The strategies are not informed by fire: fire is made to conform to other formulas" (139). This is also Scott's (1998) critique of "seeing like a state." "The proper core of fire management, however, should be fire, fire in the quotidian of routine life on the land. Mostly, we scrutinise fire on the fringe, but from perspectives on the outside looking in toward the fire; we need also to look outward from a vantage point within and centred on fire. The practical and conceptual core should be fire on the land" (Pyne 2020, 139–40).

14 I thank John Dupré for putting this point to me.

15 Alternatively, we can arbitrarily cut off the regresses, but this is to court disaster as discussed in the previous chapter on pesticides and pest control.

16 One of Verran's points is that Aboriginal and scientific fire belong to two "incommensurable" paradigms (2002, 748), but she emphasizes an epistemic or metaphysical incommensurability, hence her interest in the Aboriginal concern with place-names which makes no sense in science. The incommensurability that interests me more is at the level of patterns of action, foregrounding or backgrounding nonhuman agency. For a reasoned argument that traditional and Indigenous knowledge can contribute positively to science (and heated responses against that idea), see Nicholas (2018, 2019). For an instance of the practical and theoretical incommensurability of traditional and scientific knowledge, see chapter 8 on animism.

17 For more on these sorts of scientific measurement, see Lynch (1985) and Latour (1999), both of whom are concerned with the move from field to text in field research — in my terms, another version of the scientific detour away from the thing itself.

18 Speaking of Australia, Pyne (2020) notes that "too much of prescribed fire is a set-piece, with its conditions established bureaucratically, a designated time and place, equipment for control on hand, and so forth. If any item on the checklist is not satisfied, the process can shut down" (142). Likewise, in the United States, "prescribed fire as a set piece has become trickier, more expensive, and less agile, burdened with an ever-expanding checklist of pre-burn requirements" (7).

19 Though Aboriginal beliefs departed from my analysis in specifics — their understanding of the spirits as volitional entities, for example, that could both aid and punish right and wrong behavior. "If a thing exists it has a soul, so it can choose to help or hinder. This is why ritual matters. Correct ritual persuades a soul or spirit to help" (Gammage 2011, 127). "A healthy respect for the power of spirit forces is learned from early childhood, particularly in relation to religious or social taboos. These spiritual forces are believed to have the power to make rain, foster natural growth, assist in hunting and food gathering and even to the finding of spouses or partners. It is also believed that they have the power to act against the wishes of people if the correct ceremonies and/or rituals are not

practised or observed. And it is believed that crossing the boundaries of social taboos will also incur their wrath" (Paulson 2020, 3). We can return to the volitional aspect of animism in the following chapter.

20 Gammage brings home the importance of land care to Aborigines with a historical story: in 1830, "Mannalargenna … knew soldiers were hunting, for the Black Line, the military cordon bent on capturing every surviving Tasmanian, was under way. He knew too that smoke would betray his small band, yet still they fired the land, in the face of death toiling to do what perhaps ten times as many would once have done. Nothing shows so powerfully how crucial land care was. This was no casual burning. It was a mortal duty, a levy on the souls of brave men and women" (Gammage 2011, 137–38).

21 See also chapter 6, note 36, of this work. I thank Joel Krueger, Giovanna Colombetti, and John Gava for conversations and correspondence that have helped me see the confusions that can arise here.

22 For an extended exploration of skill acquisition in a very different context, see Rappert (2022).

23 "In the interior spinifex region Aboriginal burning is being documented by a new generation of anthropologists as testimony to the resilient power of firestick farming. In the Northern Territory Aboriginal fire practices are co-evolving into modern hybrids; controlled burning is widespread" (Pyne 2020, 6). Re Aborigines: "That past is gone. But some elements, good and bad, persist, and there are others worth hauling into the future. If there is no going back, the issue remains what to carry with us as we move on" (36). Pyne discusses an "aboriginal fire regime" presently under development in Canada, and remarks that "it's a model Australia might do well to ponder" (143). There are many angles to be thought through here. Like natural farming, Aboriginal fire practices held Australia in a condition far from equilibrium, and since they were abandoned, animals, vegetation, and the soil itself have relaxed into different configurations. Parkland has turned into scrub and forests, entire species have become extinct, and so on (Gammage 2011, chapter 1; see also, for example, Gammage 2011, 103–9, 313–15; see Cronon 1983 for similar transformations in North America). A return to Aboriginal practices would thus require, at minimum, the hard and uncertain work of restoring organisms and the land to a prior condition, just as Fukuoka had to find a transition regime for his farm. At the same time, understandings of property rights are not just different between Aboriginals and settlers but incommensurable (as touched on briefly below).

24 I developed the notion of interactive stabilization of multiple elements in the very different context of the history of particle physics in Pickering (1984a).

8. SPIRITS

Epigraph 1. Carlos Castaneda, *Tales of Power*, 24.

Epigraph 2. Davi Kopenawa, *The Falling Sky*, 313.

Epigraph 3. Lynn White Jr., "The Historical Roots of Our Ecological Crisis," 1205.

Epigraph 4. Amitav Ghosh, *The Nutmeg's Curse*, 229.

1 See also Pickering (1995a, 242–46; 2010, 73–76 passim; 2017a). For another perspective on technology and animism, see Jensen and Blok (2013).

2 I thank Casper Bruun Jensen for bringing this book and the quotation at the beginning of the following section to my attention.

3 The following quotations illustrate the range of the *xapiri*'s powers: "I was able to hear and see their singing" (Kopenawa and Albert 2013, 30). "*Omama* created the *xapiri* so we could take revenge on disease" (30). *Omama* told his son, "With these spirits, you will protect human beings and their children. . . . Don't let the evil beings and jaguars come devour them. Prevent the snakes from biting them and the scorpions from stinging them. Divert the *xawara* epidemic smoke from them. Also protect the forest. Prevent the river waters from flooding it and the rains from mercilessly drenching it. . . . Hold up the sky so it does not fall apart" (32). "From these heights [the *xapiri*] see the entire forest in the distance and warn us of the evil threatening us" (64). But also. "If we treat them [*xapiri*] badly, they can also be very aggressive and kill us. This is why we sometimes fear them" (56).

4 I have been dogged by references to witchcraft throughout my career in science studies. A standard refutation of suggestions that scientific knowledge is somehow "constructed" (Pickering 1984a) is to ask whether one believes in witches.

5 On initiation and the disciplines of shamanism, see Kopenawa and Albert (2013, 75–96). *Yakoana* powder is obtained from the *yakoana hi* tree and is mixed with the leaves and bark of other plants before being inhaled, which Kopenawa for some reason calls "drinking." It is the key to joining the spirit world: "By drinking it, you will hear [the *xapiri*'s] clamor and you too will become a spirit" (31). The parallels to, and even identity with, the better-known Amazonian drug ayahuasca are evident, though ayahuasca really is drunk in a liquid brew (see Shanon 2002). Kopenawa and Albert say that *yakoana* contains the same hallucinogenic alkaloid, DMT, as ayahuasca (492n15).

6 Kopenawa and Albert (2013, 88) refer to the shaman's transit into the spirit-realm as becoming "other" and becoming a "ghost" or "spirit." If they are on good terms, the *xapiri* aid and act on behalf of the shaman, though it seems that the shaman himself also possesses new powers in the spirit world, in healing and so on. Perhaps the most striking shamanic power is that of regurgitating "sorcery substances" and "evil objects" implanted by other shamans in sufferers' bodies (53, 54, 65, 119, 122–24).

7   Interestingly, William James, one of the founders of pragmatist philosophy, argued strongly against the virtues of sobriety: "Sobriety diminishes, discriminates, and says no; drunkenness expands, unites, and says yes. . . . It brings its votary from the chill periphery of things to the radiant core. It makes him for the moment one with truth" (James 1902, 387).

8   This chapter, intersects with a lively debate in anthropology about the "ontological turn" (Pickering 2017a). It is notable that anthropologists are, in fact, very reluctant to take animism seriously (Fontein 2021). In 2014, the anthropological journal *HAU* (vol. 4, no. 2) published a book symposium on *The Falling Sky* with contributions from six authors (and a response from Bruce Albert) and only a single essay devotes any space (a couple of paragraphs) to Kopenawa's commerce with spirits, on which it equivocates:

> [Kopenawa] describes the source of his message in remarkable detail. It is the presentation dance of the *xapiri* spirits. I have never drunk the food of the *xapiri* spirits, the hallucinogen *yakoana*, nor seen their presentation dance, so I simply do not know. . . . At this point, things get a little tricky . . . for me as a professional anthropologist, committed to translation between lived worlds. On the one hand, it would be nice to take Kopenawa at face value here, but on the other hand, I know that I don't. . . . Here, well-meaning anthropologists start resorting to get-out clauses like "metaphor." This is the small print of anthropology. Kopenawa is not talking about that, but about something very different. (Gow 2014, 305)

> Chris Vasantkumar's attempt to reinvigorate ontological discussions in anthropology begins by remarking that "the Ontological Turn is dead. Or if not dead, quite a lot of air as been let out of its tires" (2022, 819). His attempt at reinflation draws on my work and that of Judith Farquhar (2012) to argue for the "ontic capaciousness" of nature — its potential to sustain different ontologies in different practices. My examples above concerned particle physics and shamanism; Vasantkumar compares different approaches to navigation in Tahiti and the West, and the different human bodies that appear in Western biomedicine and traditional Chinese medicine. For further anthropological discussions of animism, see Qu (2021), Arumugam (2023), and Ingold (2023). Adrian Franklin (2023b) includes a nice discussion of the prevalence of animism in premodern Britain and its subsequent suppression and marginalization. I might as well add that the world is also ontically capacious enough to have sustained two quite different and incommensurable ontologies in the history of particle physics (Pickering 1984a), though neither animism nor nonmodern agency come into that story.

9   The same contrast is evident in the inner transformations associated with science and shamanism. Foucault (1988) discussed technologies of self-control like Stoicism and Christianity that fortify a dualist inner self against the outside world and that science depends on. *Yakoana*, instead, is a technology of abandon-

ment, opening the self to the influence of the other, the *xapiri*. On abandonment and nonmodern selves, see Gomart and Hennion (1999), and James (1902).

10 We could, for example, think about nondualist subject positions that are not animist. Arne Naess's deep ecology hinges on human subjects experiencing themselves as on the same plane as the rest of nature (Naess 1973). Naess himself does not suggest methods of achieving this position, but see Joanna Macy's "The Council of All Beings" (Macy 2005; Seed and Macy 2007). Antonio Carvalho has explored Buddhist technologies of the self pursued by followers of Thich Nhat Hanh that foreground interconnections with the environment (2014, 2017).

11 "*Yakoanari* violently struck me on the back of the neck and sent me backward onto the ground. I instantly lost consciousness. . . . This lasted a very long time. The *yakoana* had really made me die!" (Kopenawa and Albert 2013, 78). "If you call the spirits in vain, they accuse you of having a bitter taste and reproach you for not making them dance. Exasperated, they spit on you and cover you with ashes, then run away. When this happens to a young apprentice shaman, he begins to waste away. He soon becomes very thin and ugly. Instead of transforming into a spirit, he exposes himself to the risk of dying" (91).

12 The *xapiri* are "truly defenders of the forest" and its "true owners." Shamans intercede with the *xapiri*: "By making them come down and dance, our elders have always protected all of 'nature'" (Kopenawa and Albert 2013, 398). Kopenawa also discusses shamanic interventions in repairing environmental damage: "My father-in-law and I have often made the image of *Omama* dance to tear up and then renovate our land made sick by the gold prospectors" (391). More generally, "*Omama's* image tells us: 'Open your gardens without making them go too far. Cut up wood of fallen trunks for the fires to warm you and cook your food. Do not cut the trees just to eat their fruit. Do not damage the forest for no reason. Once it is destroyed, no other will replace it! Its richness will escape forever and you will not be able to live on this land anymore'" (382).

13 "Without [the *xapiri*], without 'ecology,' the land gets warmer and the epidemics and the evil beings get closer. . . . The spirits have always . . . prevented the earth from turning into chaos and the sky from falling"' (Kopenawa and Albert 2103, 394, 397).

14 This argument parallels the discussion of the Kariba Dam and the river god Nyaminyami at the end of chapter 5. In more general terms, Latour (1993, 2004) argues similarly that recognizing nondualist couplings with nature would slow down humanity's headlong transformations of the environment, but Latour shows no interest in nondualist alternatives to modernity such as animism (Pickering 2009a).

Epigraph 1. Robert Worth, "The Dark Reality Behind Saudi Arabia's Utopian Dreams," 10.

Epigraph 2. Arturo Escobar, "Sustaining the Pluriverse," 11.

1 Bill Gates (2021) summarizes what he calculates is needed to control global warming, most prominently the invention of new technologies of enframing.

2 For critiques of the word *Anthropocene* and suggestions for more appropriate names, see Thornton and Thornton (2015); Arons (2023); Barua, Martin, and Achtnich (2023); Lezak (2024). Donna Haraway (2016) calls her imagined sympoietic alternative regime to the Anthropocene the "Chthulucene," taking her inspiration from a spider, *Pimoa cthulhu*. She rejects any association to H. P. Lovecraft's fictional elder god, Cthulhu, though I have suggested here that reference to inscrutable animist deities can help us think about the dark side of technological systems. The latter perspective extends the reference of Chthulucene from acting-with to encompass the enframing paradigm as well as poiesis.

3 In his influential writings on the parliament of things and the politics of nature, Bruno Latour (1993, 2004) adopts a split-level view in which the level of politics and representation sits above the level of worldly practices (Pickering 2009a). He is content with modernist practices as they are, while recommending a reform of the political stratum in line with the perspective of actor-network theory. In contrast, this book is about transformations in worldly practices.

4 This continues the discussion at the end of Pickering (2010).

5 One model for such interdisciplinary centers would be the famous postwar Macy conferences on cybernetics (Pickering 2015; Pangaro 2021).

6 For a series of examples of the sorts of teaching initiatives I have in mind, see Svabo, Shanks, Carleton, and Zhou (forthcoming) on a 'creative pragmatics' approach to STEM education. I thank Connie Svabo for encouraging me to think about the form as well as the content of education. For an earlier insightful discussion of studio-based and problem-based education from a cybernetic perspective, see Glanville (2002a, 2002b). My university organizes week-long summer classes on "Grand Challenges," explicitly oriented to wicked problems, including "climate and environment emergency challenges" (University of Exeter 2024).

References

Acts of John (n.d.) The Gnostic Society Library. http://www.gnosis.org/library/grs-mead/grsm_hymnofjesus.htm.

Aggarwal, Partap. n.d. "Natural Farming Succeeds in Indian Village." *Satavic Farms.* http://satavic.org/natural-farming-succeeds-in-indian-village/. Accessed July 1, 2024.

Arons, Wendy. 2023. "We Should Be Talking about the Capitalocene." *The Drama Review* 67(1): 35–40.

Aronson, Jeffe. 1983. "Grand Canyon High Water, 1983: Grand Canyon Damn Nearly Bursts," YouTube video. Accessed July 3, 2024. https://www.youtube.com/watch?v=VPcrccxcNsI.

Arumugam, Indira. 2023. "The Sacred Unbound: Insufficient Rituals, Excess Life, and Divine Agency in Rural Tamil Nadu." HAU: *Journal of Ethnographic Theory* 13(1): 53–67.

Ashby, W. Ross. 1945. "Effect of Controls on Stability." *Nature* 155: 242–43.

Ashby, W. Ross. 1960. *Design for a Brain: The Origin of Adaptive Behaviour.* 2nd ed. London: Chapman and Hall.

Asplen, Lisa. 2008. "Going with the Flow: Living the Mangle in Environmental Management Practice." In *The Mangle in Practice: Science, Society and Becoming,* edited by Andrew Pickering and Keith Guzik, 163–84. Durham, NC: Duke University Press.

Ayre, Tina. 2020. "A Year on Our Farm — September." *Focus on Thorverton,* September 2020: 16. https://thorvertonfocus.wordpress.com/2020/09/.

Barad, Karen. 1998. "Getting Real: Technoscientific Practices and the Materialization of Reality." *differences: A Journal of Feminist Cultural Studies* 10(4): 87.

Barad, Karen. 2007. *Meeting the Universe Halfway: Quantum Physics and the Entanglement of Matter and Meaning.* Durham, NC: Duke University Press.

Barua, Maan, Rebeca Ibáñez Martín, and Marthe Achtnich. 2023. "Introduction: Plantationocene." *Fieldsights,* January 2024. https://culanth.org/fieldsights/introduction-plantationocene. Accessed July 10, 2024.

Bateson, Gregory. (1959) 2000. "Minimal Requirements for a Theory of Schizophrenia." Second Annual Albert D. Lasker Lecture, reprinted in Gregory Bateson, *Steps to an Ecology of Mind,* 2nd ed., 244–70. Chicago: University of Chicago Press.

Bateson, Gregory. 1968. "Conscious Purpose Versus Nature." In *To Free a Generation: The Dialectics of Liberation*, edited by David Cooper, 34–49. New York: Collier.

Bateson, Gregory. 2000a. *Steps to an Ecology of Mind*. 2nd ed. Chicago: University of Chicago Press.

Bateson, Gregory. 2000b. "The Roots of Ecological Crisis." In Bateson, *Steps to an Ecology of Mind*, 2nd ed., 496–501. Chicago: University of Chicago Press.

Bateson, Gregory. 2002. *Mind and Nature: A Necessary Unity*. Cresskill, NJ: Hampton Press.

Bateson, Mary C. 2005. *Our Own Metaphor*. 2nd ed. Cresskill, NJ: Hampton Press.

BBC News. 2019. "Whaley Bridge Dam Wall in Pictures." August 6, 2019. https://www.bbc.co.uk/news/uk-england-derbyshire-49199505.

Beer, Stafford. 1959. *Cybernetics and Management*. London: English Universities Press.

Beer, Stafford. 1969. "The Aborting Corporate Plan." In *Perspectives on Planning: Proceedings of the O.E.C.D. Working Symposium on Long-Range Forecasting and Planning, Bellagio, Italy 27th October–2nd November 1968*, edited by Erich Jantsch, 2–21. Paris: OECD.

Bennett, Jane. 2010. "The Agency of Assemblages and the North American Blackout." *Public Culture* 17(3): 445–65.

Berkes, Fikret, Johan Colding, and Carl Folke. 2000. "Rediscovery of Traditional Ecological Knowledge as Adaptive Management." *Ecological Applications* 10(5): 1251–62.

Bijker, Wiebe. 2002. "The Oosterschelde Storm Surge Barrier: A Test Case for Dutch Water Technology, Management, and Politics." *Technology and Culture* 43(3): 569–84.

Bijker, Wiebe. 2007. "Dike and Dams, Thick with Politics." *Isis* 98(1): 109–23.

Blair, Tony, and William Hague. 2023. *A New National Purpose: Innovation Can Power the Future of Britain*. Tony Blair Institute for Global Change, February 22, 2023. https://institute.global/sites/default/files/articles/A-New-National-Purpose-Innovation-Can-Power-the-Future-of-Britain.pdf.

Bonneuil, Christophe. 2016. "Pure Lines as Industrial Simulacra: A Cultural History of Genetics from Darwin to Johannsen." In *Heredity Explored: Between Public Domain and Experimental Science, 1850–1930*, edited by Staffan Müeller-Wille and Christina Brandt, 213–42. Cambridge, MA: MIT Press.

Bosch, Ockie, A. Ross, and Robert Benton. 2003. "Integrating Science and Management through Collaborative Learning and Better Information Management." *Systems Research and Behavioral Science* 20(2): 107–18.

Brown, Claire. 2021. "Attack of the Superweeds." *New York Times*, August 18, 2021. www.nytimes.com/2021/08/18/magazine/superweeds-monsanto.html.

Callon, Michei, Pierre Lascoumes, and Yannick Barthe. 2009. *Acting in an Uncertain World: An Essay on Technical Democracy*. Cambridge, MA: MIT Press.

Carson, Rachel. 1962. *Silent Spring*. New York: Houghton Mifflin.

Carvalho, Antonio. 2014. "Subjectivity, Ecology and Meditation: Performing Interconnectedness." *Subjectivity* 7(2): 131–50.

Carvalho, Antonio. 2017. "Ecologies of the Self in Practice — Meditation, Affect and Ecosophy." *Geografiska Annaler: Series B, Human Geography* 99(2): 207–22.

Casteneda, Carlos. 1974. *Tales of Power*. Boston: Penguin Books.

Castillo, Elizabeth. 2023. "Elwha River Transformed 10 Years After Dam Removal." OPB, April 14, 2023. https://www.opb.org/article/2022/08/02/elwha-river-transformed-10-years-after-dam-removal/.

Cheramie, Kristi. 2011. "The Scale of Nature: Modeling the Mississippi River." *Places Journal*, March 2011. placesjournal.org/article/the-scale-of-nature-modeling-the-mississippi-river/.

Chuang Tzu. 1964. *Basic Writings*. Translated by Burton Watson. New York: Columbia University Press.

Clark, Andy. 2001. *Mindware: An Introduction to the Philosophy of Cognitive Science*. Oxford: Oxford University Press.

Coastal Protection and Restoration Authority. 2020. "Mississippi River Mid-Basin Sediment Diversion Program." coastal.la.gov/our-work/key-initiatives/diversion-program/.

Coole, Diane, and Samantha Frost, eds. 2010. *New Materialisms: Ontology, Agency, and Politics*. Durham, NC: Duke University Press.

Coppin, Dawn. 2002. "Capitalist Pigs: Large-Scale Swine Facilities and the Mutual Construction of Nature and Society." PhD diss., University of Illinois at Urbana, Champaign.

Coppin, Dawn. 2003. "Foucauldian Hog Futures: The Birth of Mega-Hog Farms." *Sociological Quarterly* 44(4): 597–616.

Coppin, Dawn. 2008. "Crate and Mangle: Questions of Agency in Confinement Livestock Facilities." In *The Mangle in Practice: Science, Society and Becoming*, edited by Andrew Pickering and Keith Guzik, 46–66. Durham, NC: Duke University Press.

Cowan, Jill. 2020. "Alarmed by Scope of Wildfires, Officials Turn to Native Americans for Help." *New York Times*, October 7, 2020. https://nyti.ms/2SAVLwK.

Cronon, William. 1983. *Changes in the Land: Indians, Colonists, and the Ecology of New England*. New York: Hill and Wang.

Coulice, Ben, and Emma Murphy 2015. "White Australia's Burning Issue — What's Wrong with Bill Gammage's Book." Book Review. *Chain Reaction*, no. 123, 48–49.

Cox, Christopher. 2023. "The Trillion-Gallon Question." *New York Times*, June 22, 2023. https://www.nytimes.com/2023/06/22/magazine/california-dams.html.

Crawford, Martin. 2010. *Creating a Forest Garden: Working with Nature to Grow Edible Crops*. Totnes, UK: Green Books.

Cuntz, Michael. 2014. "Places Proper and Attached *or* the Agency of the Ground and the Collectives of Domestication." *Zeitschrift für Medien- und Kulturforschung* 5(1): 101–20.

Cussins, Charis. 1996. "Ontological Choreography: Agency through Objectification in Infertility Clinics." *Social Studies of Science* 26(3): 575–610.

de la Cadena, Marisol, and Mario Blaser, eds. 2018. *A World of Many Worlds*. Durham, NC: Duke University Press.

Dreyfus, Hubert. 2006. "Heidegger on the Connection Between Nihilism, Art, Technology, and Politics." In *The Cambridge Companion to Heidegger*, edited by Charles Guignon, 345–72. Cambridge: Cambridge University Press.

Erdbrink, Thomas. 2021. "To Avoid River Flooding: Go with the Flow, the Dutch Say." *New York Times*, September 7, 2021. https://www.nytimes.com/2021/09/07/world/europe/dutch-rivers-flood-control.html.

Escobar, Arturo. 2017. "Sustaining the Pluriverse: The Political Ontology of Territorial Struggles in Latin America." In *The Anthropology of Sustainability*, edited by Marc Brightman and Jerome Lewis, 237–56. London: Palgrave.

Fantin, Shaneen. 2013. "The Biggest Estate on Earth: How Aborigines Made Australia." *ArchitectureAU*, December 13, 2013. https://architectureau.com/articles/the-biggest-estate-on-earth-how-aborigines-made-australia/.

Farming News. 2014. "20 Per Cent of World's CO_2 From Ploughing — Soil Scientist." January 15, 2014. https://farming.co.uk/news/20-per-cent-of-world%E2%80%99s-co2-from-ploughing-%E2%80%93-soil-scientist.

Farquhar, Judith. 2012. "Knowledge in Translation: Global Science, Local Things." In *Medicine and the Politics of Knowledge*, edited by Susan Levine. Capetown, South Africa: HSRC Press.

Fatherree, Ben. 2004. *The First 75 Years: History of Hydraulics Engineering at the Waterways Experiment Station*. Vicksburg, MS: US Army Engineer Research and Development Center.

Feyerabend, Paul. 1978. *Science in a Free Society*. London: New Left Books.

Fisher, Mark. 2019. "Drifting from Rewilding." *Rewilding Earth*, March 29, 2019. https://rewilding.org/drifting-from-rewilding/.

Flavelle, Christopher. 2022a. "Here's Where the U.S. Is Testing a New Response to Rising Seas." *New York Times*, November 2, 2022. https://www.nytimes.com/2022/11/02/climate/native-americans-relocate-climate-change.html.

Flavelle, Christopher. 2022b. "In a First, U.S. Pays Tribes to Move Away from Climate Threats." *New York Times*, November 4, 2022. https://www.nytimes.com/2022/11/04/climate/native-americans-relocate-climate-change.html.

Flavelle, Christopher. 2022c. "U.S. to Pay Millions to Move Tribes Threatened by Climate Change." *New York Times*, November 30, 2022. https://www.nytimes.com/2022/11/30/climate/native-tribes-relocate-climate.html.

Fleck, Ludwik. 1979. *Genesis and Development of a Scientific Fact*. Chicago: University of Chicago Press.

Fontein, Joost. 2021. "From 'Other Worlds' and 'Multiple Ontologies' to 'a Methodological Project That Poses Questions to Solve Epistemological Problems.' What Happened to *Thinking Through Things*?" *Ethnos* 86(1): 173–88.

Foucault, Michel. 1979. *Discipline and Punish: The Birth of the Prison*. New York: Vintage Books.

Foucault, Michel. 1988. *Technologies of the Self: A Seminar with Michel Foucault*. Edited by Luther Martin, Huck Gutman, and Patrick Hutton. Amherst: University of Massachusetts Press.

Franklin, Adrian. 2008. "A Choreography of Fire: A Posthumanist Account of Australians and Eucalypts." In *The Mangle in Practice: Science, Society and Becoming*, edited by Andrew Pickering and Keith Guzik, 17–45. Durham, NC: Duke University Press.

Franklin, Adrian, ed. 2023a. *The Routledge International Handbook of More-than-Human Studies*. London: Routledge.

Franklin, Adrian. 2023b. "The Separation." In *The Routledge International Handbook of More-than-Human Studies*, edited by Adrian Franklin, 1–28. London: Routledge, 2023.

Fukuoka, Masanobu. 1978. *The One-Straw Revolution: An Introduction to Natural Farming*. New York: New York Review Books.

Fukuoka, Masanobu. 1994. *The Natural Way of Farming: The Theory and Practice of Green Philosophy*. Madras, India: Bookventure.

Fukuoka, Masanobu. 2012. *Sowing Seeds in the Desert: Natural Farming, Global Restoration, and Ultimate Food Security*. White River Junction, VT: Chelsea Green Publishing.

Fuller, T. 2020. "Reducing Fire, and Cutting Carbon Emissions, the Aboriginal Way." *New York Times*, January 17, 2020. https://www.nytimes.com/2020/01/16/world/australia/aboriginal-fire-management.html

Gammage, Bill. 2011. *The Biggest Estate on Earth: How Aborigines Made Australia*. London: Allen and Unwin.

Gates, Bill. 2021. *How to Avoid a Climate Disaster: The Solutions We Have and the Breakthroughs We Need*. New York: Knopf.

GCDAMP Wiki 2022. "A 2021 Fall HFE." http://gcdamp.com/index.php/A_2021_Fall_HFE.

GCDAMP Wiki. 2023. "High-Flow Experimental (HFE) Releases." http://gcdamp.com/index.php/The_HFE_Page.

Ghosh, Amitav. 2021. *The Nutmeg's Curse: Parables for a Planet in Crisis*. London: John Murray.

Giedion, Siegfried. 1948a. "Mechanization and Death: Meat." In *Mechanization Takes Command: A Contribution to an Anonymous History*, 209–46. New York: Oxford University Press.

Giedion, Siegfried. 1948b. "Mechanization and the Soil: Agriculture." In *Mechanization Takes Command: A Contribution to an Anonymous History*, 130–68. New York: Oxford University Press.

Gies, Erica. 2022. *Water Always Wins: Thriving in an Age of Drought and Deluge*. Chicago: University of Chicago Press.

Gilbert, Scott. 2023. "A Sympoietic View if Life: Gaia as a Holobiont Community." Preprint posted September 18, 2023. https://www.preprints.org/manuscript/202309.1072/v1.

Glanville, Ranulph. 2002a. "A (Cybernetic) Musing: Cybernetics and Human Knowing." *Cybernetics and Human Knowing* 9(1): 75–82.

Glanville, Ranulph. 2002b. "A (Cybernetic) Musing: Some Examples of Cybernetically Informed Educational Practice." *Cybernetics and Human Knowing* 9(3–4): 117–26.

Gleick, James. 1987. *Chaos: Making a New Science*. New York: Penguin.

Gomart, Emily, and Antoine Hennion. 1999. "A Sociology of Attachment: Music Amateurs, Drug Users." In *Actor Network Theory and After*, edited by John Law and John Hassard, 220–47. Oxford: Blackwell.

Gow, Peter. 2014. "'Listen to Me, Listen to Me, Listen to Me, Listen to Me . . .': A Brief Commentary on *The Falling Sky* by Davi Kopenawa and Bruce Albert." HAU: *Journal of Ethnographic Theory* 4(2): 301–9.

Grams, P. E., J. C. Schmidt, S. A. Wright, D. J. Topping, T. S. Melis, and D. M. Rubin. 2015. "Building Sandbars in the Grand Canyon." EOS: *Earth and Space Science News*, June 3, 2015. https://eos.org/features/building-sandbars-in-the-grand -canyon.

Gray, Kathleen, and Julie Bosman. 2020. "As Virus Lingers in Michigan a New Crisis Arrives: Flooding." *New York Times*. https://www.nytimes.com/2020/05/20/us /michigan-flooding-dams-midland.html.

Green, Ronald. n.d. "Farming Satori: Zen and the Naturalist Farmer: Fukuoka Masanobu." https://terebess.hu/zen/mesterek/Farming_Satori.pdf.

Gunderson, Lance, and Stephen Light. 2006. "Adaptive Management and Adaptive Governance in the Everglades Ecosystem." *Policy Science* 39: 323–34.

Hajek, Ann, and Jorgen Eilenberg. 2018. *Natural Enemies: An Introduction to Biological Control*. 2nd ed. Cambridge: Cambridge University Press.

Hanson, Norwood R. 1958. *Patterns of Discovery: An Inquiry into the Conceptual Foundations of Knowledge*. Cambridge: Cambridge University Press.

Haraway, Donna. 2003. *The Companion Species Manifesto: Dogs, People, and Significant Otherness*. Chicago: Prickly Paradigm Press.

Haraway, Donna. 2004. *The Haraway Reader*. New York: Routledge.

Haraway, Donna. 2008. *When Species Meet*. Minneapolis: University of Minnesota Press.

Haraway, Donna. 2016. *Staying with the Trouble: Making Kin in the Chthulucene*. Durham, NC: Duke University Press.

Harries-Jones, Peter. 1995. *A Recursive Vision: Ecological Understanding and Gregory Bateson*. Toronto: University of Toronto Press.

Haudricourt, André. 1969. "Domestication of Animals, Cultivation of Plants and Human Relations." *Social Science Information* 8(3): 163–72.

Heidegger, Martin. (1954) 1977. "The Question Concerning Technology." In *Martin Heidegger: Basic Writings*, edited by David Krell, 217–38. Abingdon, UK: Routledge.

Heidegger, Martin. 1981. "'Only a God Can Save Us': The *Spiegel* Interview (1966)." In *Heidegger: The Man and the Thinker*, edited by Thomas Sheehan, 45–67. Chicago: Precedent Publishing.

Hinchliffe, Steve. 2022. "Postcolonial Global Health, Post-Colony Microbes and Antimicrobial Resistance." *Theory, Culture and Society* 39(3): 145–68.

Hinchliffe, Steve, and Kim Ward. 2014. "Geographies of Folded Life: How Immunity Reframes Biosecurity." *Geoforum* 53: 136–44.

Holling, C. S. 1973. "Resilience and Stability of Ecological Systems." *Annual Review of Ecological Systems* 4: 1–23.

Holling, C. S., ed. 1978. *Adaptive Environmental Assessment and Management*. London: Wiley.

Holling, C. S., and Gary Meffe. 1996. "Command and Control and the Pathology of Natural Resource Management." *Conservation Biology* 10(2): 328–37.

Holling, C. S., and Shana Sundstrom. 2015. "Adaptive Management, A Personal History." In *Adaptive Management of Social-Ecological Systems*, edited by Craig R. Allen and Ahjond S. Garmestani, 11–25. Dordrecht: Springer.

Horkheimer, Max, and Theodor Adorno. 1979. *Dialectic of Enlightenment*. London: Verso.

Horton, D. R. 1982. "The Burning Question: Aborigines, Fire and Australian Ecosystems." *Mankind* 13(3): 237–63.

Hutchins, Edwin. 1995. *Cognition in the Wild*. Cambridge, MA: MIT Press.

Huxley, Aldous. 1954. *The Doors of Perception*. New York: Harper.

Ideas. n.d. "Implementing the Fukuoka's Natural Farming Methodologies." http://www.ideasonline.org/public/pdf/FukuocaRassegnaENG2.pdf.

Ingold, Tim. 2023. "A Circumpolar Night's Dream." In *The Routledge International Handbook of More-than-Human Studies*, edited by Adrian Franklin, 79–100. London: Routledge.

James, William. 1902. *The Varieties of Religious Experience: A Study in Human Nature*. New York: Longmans, Green.

Jensen, Casper B. 2020. "The Anthropocene Eel: Emergent Knowledge, Ontological Politics and New Propositions for an Age of Extinctions." *Anthropocenes: Human, Inhuman, Posthuman* 1(1): 1–10.

Jensen, Casper B., and Anders Blok. 2013. "Techno-Animism in Japan: Cosmograms, Actor-Network Theory, and the Enabling Powers of Non-Human Agencies." *Theory, Culture and Society* 30(2): 84–115.

Jensen, Pablo. 2021. *Your Life in Numbers: Modeling Society Through Data*. Cham, Switzerland: Springer.

Jensen, Pablo. 2022. "Introducing Simple Models of Social Systems." *American Journal of Physics* 90(6): 462–68.

Jensen, Pablo, and F. Li Vigni, eds. Forthcoming. *Critical Studies of Complexity: Theories. Notions, Translations and Normativity*. Paris: Editions Matériologiques.

Jullien, François. 1999. *The Propensity of Things: Toward a History of Efficacy in China*. New York: Zone Books.

Juniper, Andrew. 2003. *Wabi Sabi: The Japanese Art of Impermanence*. Tokyo: Tuttle.

Keulartz, Jozef. 2012. "The Emergence of Enlightened Anthropocentrism in Ecological Restoration." *Nature and Culture* 7(1): 48–71.

Kimmelman, Michael. 2013. "Going with the Flow." *New York Times*, February 13, 2013.

Kolbert, Elizabeth. 2012. "Recall of the Wild: The Quest to Engineer a World before Humans." *New Yorker*, December 24, 2012. www.newyorker.com/magazine/2012/12/24/recall-of-the-wild.

Kolbert, Elizabeth. 2019. "Louisiana's Disappearing Coast." *New Yorker*, April 21, 2019. https://www.newyorker.com/magazine/2019/04/01/louisianas-disappearing -coast.

Konrad, C., A. Warner, and J. Higgins. 2012. "Evaluating Dam Re-Operation for Freshwater Conservation in the Sustainable Rivers Project." *River Research and Applications*, 29: 777–92.

Kopenawa, Davi, and Bruce Albert. 2013. *The Falling Sky: Words of a Yanomami Shaman.* Cambridge, MA: Harvard University Press.

Korn, Larry. 1978. "Editor's Introduction." In *The One-Straw Revolution: An Introduction to Natural Farming* by Masanobu Fukuoka, xvii–xxviii. New York: New York Review Books.

Korn, Larry. 2012. "Masanobu Fukuoka and Natural Farming." *Final Straw*, September 5, 2012. www.finalstraw.org/masanobu-fukuoka-and-natural-farming/.

Krueger, Joel. 2009. "Knowing through the Body: The *Daodejing* and Dewey." *Journal of Chinese Philosophy* 36(1): 31–52.

Kuhn, Thomas. 1962. *The Structure of Scientific Revolutions.* Chicago: University of Chicago Press.

Laozi. 2008. *Daodejing.* Translated by E. Ryden. Oxford: Oxford University Press.

Lappé, Frances. 1978. "Introduction." In Masanobu Fukuoka, *The One-Straw Revolution: An Introduction to Natural Farming*, vii–x. New York: New York Review Books.

Latour, Bruno. 1983. "Give Me a Laboratory and I Will Raise the World." In *Science Observed: Perspectives on the Social Study of Science*, edited by Karin D. Knorr-Cetina and Michael Mulkay, 141–70. Beverly Hills, CA: Sage.

Latour, Bruno. 1987. *Science in Action: How to Follow Scientists and Engineers Through Society.* Cambridge, MA: Harvard University Press.

Latour, Bruno 1993. *We Have Never Been Modern.* Cambridge, MA: Harvard University Press.

Latour, Bruno. 1999. "Circulating Reference: Sampling the Soil in the Amazon Forest." In *Pandora's Hope: Essays on the Reality of Science Studies*, 24–79. Cambridge, MA: Harvard University Press.

Latour, Bruno. 2004. *Politics of Nature: How to Bring the Sciences into Democracy.* Cambridge, MA: Harvard University Press.

Law, John. 2015. "What's Wrong with a One-World World?" *Distinktion* 16(1): 126–39.

Lee, Kai. 1991. *Compass and Gyroscope: Integrating Science and Politics for the Environment.* Washington, DC: Island Press.

Lenton, Timothy, and Bruno Latour. 2018. "Gaia 2.0: Could Humans Add Some Level of Self-Awareness to Earth's Self-Regulation?" *Science* 361 (6407): 1066–69.

Leslie, Jacques. 2014. "Large Dams Just Aren't Worth the Cost." *New York Times*, August 22, 2014. https://www.nytimes.com/2014/08/24/opinion/sunday/large-dams -just-arent-worth-the-cost.html.

Lezak, Stephen. 2024. "Scientists Just Gave Humanity an Overdue Reality Check. The World Will Be Better for It." *New York Times*, March 22, 2024. https://www .nytimes.com/2024/03/22/opinion/anthropocene-planet-geology.html.

Li, Shuhua. 2012. "Natural Philosophy of *Zhouyi* and Life Practice." *Frontiers of Philosophy in China* 7(2): 179–90.

London, Scott. 2005. "Permaculture: A Quiet Revolution — An Interview with Bill Mollison." *Green Living*, scott.london/interviews/mollison.html.

Lorimer, Jamie. 2017. "Probiotic Environmentalities: Rewilding with Wolves and Worms." *Theory, Culture and Society* 34(4): 27–48.

Lorimer, Jamie, and Clemens Driessen. 2014. "Wild Experiments at the Oostvaardersplassen: Rethinking Environmentalism in the Anthropocene." *Transactions of the Institute of British Geographers* 39(2): 169–81.

Lorimer, Jamie, Christopher James Sandom, Paul R. Jepson, Chris Doughty, Maan Barua, Keith Kirby. 2015. "Rewilding: Science, Practice, and Politics." *Annual Review of Environment and Resources* 40(1): 39–62.

Lovelock, James. 1979. *Gaia: A New Look at Life on Earth*. Oxford: Oxford University Press.

Lovelock, James. 2007. *The Revenge of Gaia: Why the Earth Is Fighting Back — And How We Can Still Save Humanity*. London: Penguin.

Lynch, Michael. 1985. "Discipline and the Material Form of Images: An Analysis of Scientific Visibility." *Social Studies of Science* 15(1): 37–66.

Macedo, Marcia, and Valéria Pereira. 2020. "We Know How to Stop the Fires." *New York Times*, October 2, 2020. https://www.nytimes.com/2020/10/02/opinion/amazon-rainforest-fire-prevention.html.

Mach, Katharine, and A. Siders 2021a. "Is Your Town Threatened by Floods or Fires? Consider a 'Managed Retreat.'" *New York Times*, July 16, 2021. https://www.nytimes.com/2021/07/16/opinion/managed-retreat-climate-change.html.

Mach, Katharine, and A. Siders. 2021b. "Reframing Strategic, Managed Retreat for Transformative Climate Adaptation." *Science* 372 (6548): 1294–99.

Macy, Joanna. 2005. "The Council of All Beings." *Encyclopedia of Religion and Nature* 1: 425–29. www.rainforestinfo.org.au/deep-eco/Joanna%20Macy.htm.

Mageau, Michael. 2016. "The Aquaponics Solution." *Solutions*, February 22, 2016. https://thesolutionsjournal.com/the-aquaponics-solution/.

Maharidge, Dale. 2023. "The Colorado River Is Running Dry, but Nobody Wants to Talk About the Mud." *New York Times*, March 21, 2023. www.nytimes.com/2023/03/20/opinion/colorado-river-lake-powell-glen-canyon-dam.html.

Maitlis, Emily, and Bernard-Henri Lévy, *Newsnight*, BBC2, July 30, 2020.

Marchese, David. 2022. "Brian Eno Reveals the Hidden Purpose of All Art." *New York Times*, November 13, 2022.

Mashal, Mujib, and Hari Kumar. 2021. "Glacier Bursts in India, Leaving More than 100 Missing in Floods." *New York Times*, February 7, 2021. https://www.nytimes.com/2021/02/07/world/asia/india-glacier-flood-uttarakhand.html.

McCarthy, Michael. 2010. "Anger in Hungary, Fear Downstream as Toxic Contamination Spreads." *Independent*. October 8, 2010. https://www.independent.co.uk/climate-change/news/anger-in-hungary-fear-downstream-as-toxic-contamination-spreads-2101071.html.

McPhee, John. 1989. *The Control of Nature*. New York: Farrar, Straus, Giroux.

McVeigh, Tracy. 2014. "The Dutch Solution: Live with Water, Don't Fight It." *Guardian*, February 16, 2014. www.theguardian.com/environment/2014/feb/16/flooding -netherlands.

Medina, Eden. 2014. *Cybernetic Revolutionaries: Technology and Politics in Allende's Chile*. Cambridge, MA: MIT Press.

Meilvang, Marie. 2021. "From Rain as Risk to Rain as Resource: Professional and Organizational Changes in Urban Rainwater Management." *Current Sociology* 69 (7): 1034–50.

Melis, Theodor, Paul E. Grams, Theodore A. Kennedy, Barbara E. Ralston, Christopher T. Robinson, John Schmidt, Lara M. Schmit, Richard Valdez, and Scott A. Wright. 2011. *Three Experimental High-Flow Releases from Glen Canyon Dam, Arizona — Effects on the Downstream Colorado River Ecosystem*. Flagstaff, AZ: US Department of the Interior, US Geological Survey. pubs.usgs.gov/fs/2011/3012/.

Meretzky, Vicky, David Wegner, and Lawrence Stevens. 2000. "Balancing Endangered Species and Ecosystems: A Case Study of Adaptive Management in Grand Canyon." *Environmental Management* 26(6): 579–86.

Mollison, Bill, and David Holmgren. 1978. *Permaculture One: A Perennial Agriculture for Human Settlements*. London: Transworld.

Monbiot, George. 2013. *Feral: Searching for Enchantment on the Frontiers of Rewilding*. London: Allen Lane.

Mother Earth News. 1987. "Ecological Farming: A Conversation with Fukuoka, Jackson and Mollison." *Mother Earth News*, March 1, 1987. https://www.motherearthnews .com/homesteading-and-livestock/ecological-farming-zmaz87mazgoe.

Mukerji, Chandra. 1994. "The Political Mobilization of Nature in Seventeenth-Century French Formal Gardens." *Theory and Society* 23(5): 651–77.

Mukerji, Chandra. 1997. *Territorial Ambitions and the Gardens of Versailles*. Cambridge: Cambridge University Press.

Mukerji, Chandra. 2002. "Material Practices of Domination: Christian Humanism, the Built Environment, and Techniques of Western Power." *Theory and Society* 31(1): 1–34.

Munoz, Samuel, Liviu Giosan, Matthew D. Therrell, Jonathan W. F. Remo, Zhixiong Shen, Richard M. Sullivan, Charlotte Wiman, Michelle O'Donnell, Jeffrey P. Donnelly. 2018. "Climatic Control of Mississippi River Flood Hazard Amplified by River Engineering." *Nature* 556: 95–98.

Naess, Arne. 1973. "The Shallow and the Deep, Long-Range Ecology Movements: A Summary." *Inquiry* 16(1–4): 95–100.

Nafziger, Emerson. 2020. "Planting Corn and Soybeans in 2020." *The Bulletin: Pest Management and Crop Development Information for Illinois*, April 7, 2020. bulletin.ipm .illinois.edu/?p=4976.

Nagourney, Adam, and Henry Fountain. 2017. "Oroville Is a Warning for California Dams, as Climate Change Adds Stress." *New York Times*, February 14, 2017, https:// www.nytimes.com/2017/oroville-dam-climate-change-california.html.

NASA Earth Observatory. 2023. "High Flow at Glen Canyon Dam." https://earth observatory.nasa.gov/images/151320/high-flow-at-glen-canyon-dam.

Neal, Timothy. 2012. Review of *"The Biggest Estate on Earth." Arena Magazine*, February 2012. arena.org.au/the-biggest-estate-on-earth-review-by-timothy-neale/.

National Resources Defense Council. 2021. "Regenerative Agriculture 101." November 29, 2021. https://www.nrdc.org/stories/regenerative-agriculture-101.

Needham, Joseph. 1956. *Science and Civilisation in China.* Vol. 2, *History of Scientific Thought.* Cambridge: Cambridge University Press.

Needham, Joseph. 1971. *Science and Civilisation in China.* Vol. 4, *Physics and Physical Technology, Part III, Civil Engineering and Nautics.* Cambridge: Cambridge University Press.

Nicholas, George. 2018. "It's Taken Thousands of Years, but Western Science Is Finally Catching Up to Traditional Knowledge." *Conversation*, February 15, 2018. https:// theconversation.com/its-taken-thousands-of-years-but-western-science-is -finally-catching-up-to-traditional-knowledge-90291.

Nicholas, George. 2019. "An Uneasy Alliance: Indigenous Traditional Knowledge Enriches Science." *Conversation*, February 18, 2019. https://theconversation.com /an-uneasy-alliance-indigenous-traditional-knowledge-enriches-science -109212.

Nijhuis, Michelle. 2014. "World's Largest Dam Removal Unleashes U.S. River After Century of Electric Production." *National Geographic*, August 27, 2014. https:// www.nationalgeographic.com/science/article/140826-elwha-river-dam -removal-salmon-science-olympic.

Olden, Julian D., Christopher P. Konrad, Theodore S. Melis, Mark J. Kennard, Mary C. Freeman, Meryl C. Mims, Erin N. Bray, et al. 2014. "Are Large-Scale Flow Experiments Informing the Science and Management of Freshwater Ecosystems?" *Frontiers of Ecology and Environment* 12(3): 176–85.

Pangaro, Paul. 2021. "#New Macy Meetings Manifesto — Conversations for Action." Last updated March 5, 2021. https://docs.google.com/document/d /1jsN3nD2eLwRr35IyPFFBKmk49uP9urZ-ofgIbGUx_p4/edit#heading=h .uew3e7r4ihot.

Pask, Gordon. 1958. "Organic Control and the Cybernetic Method." *Cybernetica*, 1(3): 155–73.

Paulson, Graham. 2020. "Aboriginal Spirituality." *Australians Together.* australianstogether.org.au/discover/indigenous-culture/aboriginal-spirituality/. Accessed July 1 2024.

Perrow, Charles. 1984. *Normal Accidents: Living with High-Risk Technologies.* New York: Basic Books.

Pickering, Andrew. 1981. "The Hunting of the Quark." *Isis* 72(2): 216–36.

Pickering, Andrew. 1984a. *Constructing Quarks: A Sociological History of Particle Physics.* Chicago: University of Chicago Press.

Pickering, Andrew. 1984b. "Against Putting the Phenomena First: The Discovery of the Weak Neutral Current." *Studies in History and Philosophy of Science* 15(2): 85–117.

Pickering, Andrew. 1993. "The Mangle of Practice: Agency and Emergence in the Sociology of Science." *American Journal of Sociology* 99(3): 559–89.

Pickering, Andrew. 1995a. *The Mangle of Practice: Time, Agency, and Science*. Chicago: University of Chicago Press.

Pickering, Andrew. 1995b. "Cyborg History and the World War II Regime." *Perspectives on Science* 3(1): 1–48.

Pickering, Andrew. 2000. "In the Thick of Things and the Politics of Becoming." Paper presented at the Thirteenth Inter-Nordic Symposium in Philosophy, Bergen, Norway, May 18–21, 2000. Reprinted in *The Routledge International Handbook of More-than-Human Studies*, edited by Adrian Franklin, 31–41. London: Routledge, 2023.

Pickering, Andrew. 2001a. "In the Thick of Things." Keynote address at the conference on Taking Nature Seriously: Citizens, Science, and Environment, University of Oregon, Eugene, Oregon, February 25–27, 2001.

Pickering, Andrew. 2001b. "Reading the *Structure*." *Perspectives on Science* 9(4): 499–510.

Pickering, Andrew. 2005a. "Asian Eels and Global Warming: A Posthumanist Perspective on Society and the Environment." *Ethics and the Environment* 10(20): 29–43.

Pickering, Andrew. 2005b. "Decentring Sociology: Synthetic Dyes and Social Theory." *Perspectives on Science* 13(3): 352–405.

Pickering, Andrew. 2008. "New Ontologies." In *The Mangle in Practice: Science, Society and Becoming*, edited by Andrew Pickering and Keith Guzik, 1–14. Durham, NC: Duke University Press.

Pickering, Andrew. 2009a. "The Politics of Theory: Producing Another World, with Some Thoughts on Latour." *Journal of Cultural Economy* 2 (1–2): 199–214.

Pickering, Andrew. 2009b. "Beyond Design: Cybernetics, Biological Computers and Hylozoism." *Synthese* 168(3): 469–91.

Pickering, Andrew. 2010. *The Cybernetic Brain: Sketches of Another Future*. Chicago: University of Chicago Press.

Pickering, Andrew. 2011. "Cyborg Spirituality." *Medical History* 55(3): 349–53.

Pickering, Andrew. 2013a. "Being in an Environment: A Performative Perspective." *Natures Sciences Sociétés* 21(1): 77–83.

Pickering, Andrew. 2013b. "Ontology and Antidisciplinarity." In *Interdisciplinarity: Reconfigurations of the Social and Natural Sciences*, edited by Andrew Barry and Georgina Born, 209–25. London: Routledge.

Pickering, Andrew. 2014. "Islands of Stability: From Cellular Automata to the Occupy Movement." *Zeitschrift für Medien- und Kulturforschung* 14(1): 121–34.

Pickering, Andrew. 2015. "'The Next Macy Conference: A New Interdisciplinary Synthesis." *Technology and Society* 34 (5): 37–38.

Pickering, Andrew. 2016. "Art, Science and Experiment." *MaHKUscript: Journal of Fine Art Research* 1(1): 1–6.

Pickering, Andrew. 2017a. "The Ontological Turn: Taking Different Worlds Seriously." *Social Analysis*, 61(2): 134–50.

Pickering, Andrew. 2017b. "In Our Place: Performance, Dualism, and Islands of Stability." *Common Knowledge* 23(3): 381–95.

Pickering, Andrew. 2019. "Techniques de l'engagement: la cybernétique et l'*Internet of Things*." *Zilsel* 1(5): 195–209.

Pickering, Andrew. 2021. "Shared Habitats and Ueküll's Bubble." In *Shared Habitats: A Cultural Inquiry into Living Spaces and Their Inhabitants*, edited by Ursula Damm and Mindaugas Gapševičius, 29–46. Bielefeld: Transcript Verlag.

Pickering, Andrew. 2023. "What Is Agency? A View from Science Studies and Cybernetics." *Biological Theory* 19(1): 16–21.

Pickering, Andrew. Forthcoming a. "Cybernetic Art." In *The Bloomsbury Encyclopaedia of New Media Art*, edited by Charlie Gere. London: Bloomsbury.

Pickering, Andrew. Forthcoming b. "Wicked Problems and the Cybernetic Method." In *Critical Studies of Complexity: Theories. Notions, Translations and Normativity*. edited by Pablo Jensen and Fabrizio Li Vigni. Editions Matériologiques. www.ixxi.fr /lyonpickeringcomplexitypolitics.pdf?lang=en.

Plumwood, Valerie. 2001. "Nature as Agency and the Prospects for a Progressive Naturalism." *Capitalism Nature Socialism* 12(4): 3–32.

Pope Francis. 2015. *Laudato si'*. Vatican City: Vatican Press.

Prigogine, Ilya, and Isabelle Stengers. 1984. *Order Out of Chaos: Man's New Dialogue with Nature*. New York: Bantam Books.

Puig de la Bellacasa, Maria. 2019. "Re-Animating Soils: Transforming Human-Soil Affections Through Science, Culture and Community." *Sociological Review Monographs* 67(2): 391–407.

Pyne, Stephen. 2020. *The Still-Burning Bush*. London: Scribe.

Qu, Feng. 2021. "Embodiment of Ancestral Spirits, the Social Interface, and Ritual Ceremonies: Construction of the Shamanic Landscape Among the Daur in North China." *Religions* 12(8): 567–85.

Rana, Surinder, and Sandeep Manuja. 2020. "Lecture 6a: Natural Farming." www .researchgate.net/publication/339310191_Lecture_6a_Natural_Farming. Accessed July 2, 2024.

Randle, Timothy J., Joseph K. Lyons, Rick J. Christensen, and Ryan D. Stephen. 2007. *Colorado River Ecosystem Sediment Augmentation Appraisal Engineering Report*. Denver, CO: US Bureau of Reclamation.

Rappert, Brian. 2022. *Performing Deception: Learning, Skill and the Art of Conjuring*. Cambridge: Open Book Publishers.

Rathore, Vaishnavi. 2019. "Going Back to the Roots with Natural Farming," *Mongabay*. https://india.mongabay.com/2019/11/going-back-to-the-roots-with-natural -farming/.

Renfro, Alisha. 2019. "5 Reasons Why 2019's Mississippi River Flood Is the Most Unprecedented of Our Time." *Delta Dispatches*, June 27, 2019. https://mississippi riverdelta.org/5-reasons-why-2019s-mississippi-river-flood-is-the-most -unprecedented-of-our-time/.

Rewilding Europe. 2012. "Rewilding as a Tool and the Role of Science." August 14, 2012. https://rewildingeurope.com/news/rewilding-as-a-tool-and-the-role-of -science/.

Rice, James. 2013. "Controlled Flooding in the Grand Canyon: Drifting Between Instrumental and Ecological Rationality in Water Management." *Organization and Environment* 26(4): 412–30.

Rich, Nathaniel. 2020."Destroying a Way of Life to Save Louisiana." *New York Times*, July 21, 2020. www.nytimes.com/interactive/2020/07/21/magazine/louisiana -coast-engineering.html.

Richter, Brian, and Gregory Thomas. 2007. "Restoring Environmental Flows by Modifying Dam Operations." *Ecology and Society*, 12(1): 12. https://www.jstor.org/stable /26267852.

Rittel, Horst, and Melvin Webber. 1973. "Dilemmas in a General Theory of Planning." *Policy Sciences* 4: 155–69.

Robbins, Jim. 2024. "To Protect Redwoods They Lit a Fire," *New York Times*, July 6, 2024. https://www.nytimes.com/2024/07/09/science/redwoods-wildfires-indigenous -tribes-california.html.

Rusho, W. L. 2008. "Bumpy Road for Glen Canyon Dam." In *The Bureau of Reclamation: Historical Essays from the Centennial Symposium, Volumes I and II*, 523–50. Denver, CO: Bureau of Reclamation.

Russell, Lynette. 2020. "Indigenous Practice of Cultural Burning and the Bushfires Prevention Conversation." *Lens* (Monash University), January 8, 2020. https:// lens.monash.edu/@lynette-russell/2020/01/08/1379433?slug=bringing -indigenous-knowledge-into-the-bushfires-conversation.

Schagrin, Winston. 2021. "Wildfires Are Intensifying: Here's Why, and What Can Be Done." *New York Times*, July 16, 2021. https://www.nytimes.com/2021/07/16 /climate/wildfires-smoke-safety-questions.html.

Schivelbusch, Wolfgang. 1986. *The Railway Journey: The Industrialization of Time and Space in the 19th Century*. Berkeley: University of California Press.

Schmidt, John, et al. 2001. "The 1996 Controlled Flood in Grand Canyon: Flow, Sediment Transport, and Geomorphic Change." *Ecological Applications* 11(2): 657–71.

Schwartz, John. 2020. "A Mini-Mississippi River May Help Save Louisiana's Vanishing Coast." *New York Times*, February 25, 2020. https://www.nytimes.com/2020/02/25 /climate/louisiana-mississippi-river-model.html.

Scott, James. 1998. *Seeing like a State: How Certain Schemes to Improve the Human Condition Have Failed*. New Haven, CT: Yale University Press.

Scott, James. 2017. *Against the Grain: A Deep History of the Earliest States*. New Haven, CT: Yale University Press.

Scott, James. 2020. "In Praise of Floods." Albert O Hirschman Prize Lecture, December 4, 2020. https://www.youtube.com/watch?v=YzUP9Mjml6Y.

Seed, John, and Joanna Macy. 2007. *Thinking like a Mountain: Towards a Council of All Beings*. Gabriola Island, Canada: New Catalyst Books.

Serpell, Namwali. 2020. "Learning from the Kariba Dam." *New York Times*, July 22, 2020. www.nytimes.com/interactive/2020/07/22/magazine/zambia-kariba-dam.html.

Severson, Kim. 2021. "No Soil. No Growing Seasons. Just Add Water and Technology." *New York Times*, July 6, 2021. https://www.nytimes.com/2021/07/06/dining/hydroponic-farming.html.

Shanon, Benny. 2002. *The Antipodes of the Mind: Charting the Phenomenology of the Ayahuasca Experience*. Oxford: Oxford University Press.

Shiva, Vandana. 1988. *Staying Alive: Women, Ecology and Survival in India*. London: Zed Books.

Slingerland, Edward. 2003. *Effortless Action: Wu-wei as Conceptual Metaphor and Spiritual Ideal in Early China*. Oxford: Oxford University Press.

Smith, Thomas. n.d. "Self-So and Self-Sow: An Exploration of the Parallels between Permaculture and Daoism." Master's thesis, University of Cork.

Sullivan, Helen. 2020. "'Australia"s Fire Season Ends, and Researchers Look to the Next One." *New York Times*, April 21, 2020. https://www.nytimes.com/2020/04/21/science/australia-wildfires-technology-drones.html.

Svabo, Connie, Michael Shanks, Tamara Carleton, and Chunfang Zhou, eds. Forthcoming. *Creative Pragmatics Through Active Learning in STEM Education*. Berlin: Springer.

Tegta, Malvika. 2010. "Masanobu Fukuoka: The Man Who Did Nothing." DNA. Last updated August 21, 2010. https://www.dnaindia.com/lifestyle/report-masanobu-fukuoka-the-man-who-did-nothing-1426864.

Tenz, Courtney, and Philip Oltermann. 2021. "'It's All Wrecked': German Town Stunned by Flood Damage." *Guardian*, July 16, 2021. https://www.theguardian.com/world/2021/jul/16/all-wrecked-german-town-stunned-flood-damage.

Theunissen, Bert. 2019. "The Oostvaardersplassen Fiasco." *Isis* 110(2): 341–45.

Thornton, Thomas, and Patricia Thornton. 2015. "The Mutable, the Mythical, and the Managerial: Raven Narratives and the Anthropocene." *Environment and Society* 6: 66–86.

Tsing, Anna. 2012. "Unruly Edges: Mushrooms as Companion Species." *Environmental Humanities* 1(1): 141–54.

Tsing, Anna. 2015. *The Mushroom at the End of the World: On the Possibility of Life in Capitalist Ruins*. Princeton, NJ: Princeton University Press.

United States Bureau of Reclamation. 2018. "Interior Initiates High-Flow Experiment at Glen Canyon Dam." *Cision*, November 16, 2018. https://www.prweb.com/releases/U_S_Department_of_the_Interior_Initiates_High_Flow_Experiment_at_Glen_Canyon_Dam/prweb15896618.htm.

van Alphen, Sander. 2020. "Room for the River: Innovation or Tradition? The Case of the Noordward." In *Adaptive Strategies for Water Heritage*, edited by Carola Hein, 308–23. Amsterdam: Springer.

Vasantkumar, Chris. 2022. "Not 'Multiple Ontologies' but Ontic Capaciousness." HAU: *Journal of Ethnographic Theory* 12(3): 819–35.

Vera, Frans. 2009. "Large-Scale Nature Development — The Oostvaardersplassen." Supplement, *British Wildlife* 20(5): 28–36.

Verran, Helen. 2002. "A Postcolonial Moment in Science Studies: Alternative Firing Regimes of Environmental Scientists and Aboriginal Landowners." *Social Studies of Science* 32(5–6): 729–62.

Viveiros de Castro, Eduardo. 2012. *Cosmological Perspectivism in Amazonia and Elsewhere: Four Lectures Given in the Department of Social Anthropology, University of Cambridge, February–March 1998.* HAU Masterclass Series, Manchester, UK. http://haubooks.org/cosmological-perspectivism-in-amazonia/.

von Foerster, Heinz. 2014. *The Beginning of Heaven and Earth Has No Name.* New York: Fordham University Press.

Watson, Julia. 2019. *Lo-TEK, Design by Radical Imagination.* Cologne: Taschen Books.

Watts, Alan. 1957. *The Way of Zen.* New York: Pantheon.

Wawrytko, Sandra. 2005. "The Viability (Dao) and Virtuosity (De) of Daoist Ecology: Reversion (Fu) as Renewal." *Journal of Chinese Philosophy* 32(1): 89–103.

White, Lynn, Jr. 1967. "The Historical Roots of Our Ecological Crisis." *Science,* no. 155: 1203–7.

Whitney, Elspeth. 2015. "Lynn White Jr.'s 'The Historical Roots of Our Ecological Crisis' After 50 Years." *History Compass* 13(8): 399–410.

Wipulasena, Aanya, and Mujib Mashal. 2021. "Sri Lanka's Plunge into Organic Farming Brings Disaster." *New York Times,* December 7, 2021. https://www.nytimes.com/2021/12/07/world/asia/sri-lanka-organic-farming-fertilizer.html.

Worth, Robert. 2021. "The Dark Reality Behind Saudi Arabia's Utopian Dreams." *New York Times,* January 28, 2021. www.nytimes.com/2021/01/28/magazine/saudi-arabia-neom-the-line.html.

Wreg, Rob. n.d. "How Does Ploughing Release Carbon? — Causes and Potential Solutions." *Innovate Eco* (blog). Accessed July 2, 2024 . innovate-eco.com/how-does-ploughing-release-carbon-causes-and-potential-solutions/.

Wynter, Coral. 2015. "What's Right with Bill Gammage's Book." *Green Left,* March 15, 2015. https://www.greenleft.org.au/content/whats-right-bill-gammages-book.

Yardley, Jim, and Laurie Goodstein. 2015. "Pope Francis, in Sweeping Encyclical, Calls for Swift Action on Climate Change." *New York Times,* June 19, 2015. https://www.nytimes.com/2015/06/19/world/europe/pope-francis-in-sweeping-encyclical-calls-for-swift-action-on-climate-change.html.

Yu, Jiyuan. 2008. "Living wih Nature: Stoicism and Daoism." *History of Philosophy Quarterly* 26(1): 1–19.

Zhang, Shanghong, Yujun Yi, Yan Liu, and Xingkui Wang. 2013. "Hydraulic Principles of the 2,268-Year-Old Dujiangyan Project in China." *Journal of Hydraulic Engineering* 139(5): 538–46.

Zwart, Hub. 2003. "Aquaphobia, Tulipmania, Biophilia: A Moral Geography of the Dutch Landscape." *Environmental Values,* 12(1): 107–28.

Index

32; Mississippi River, letting go, and, 20–28; paradigm shift from, 49–55, 100–102; performative excess of, 33; poiesis and, 12–13; religion and, 53–55, 117n10, 118n11; science and, 7–9, 26–28, 32, 38, 98, 108n7; social relations, 89–90; stance of, 1–2, 4–5, 10, 12, 14, 98; as technology, 5–6, 22–23, 32–35, 47–48, 98; term usage, 4–5; in wicked problems, 11, 99

engineering: by ACE, 3, 21–28; adaptive architecture and, x; bad, 102–3; enframing and, 51, 102–3, 110n7; in erosion control, 29–31; of Mississippi River levees, 21–28; paradigm shift in, 51–52

environmental crisis, xi, 77, 102–3; in Anthropocene, 100; climate change and, 24, 113n11; dam failures and, 112n3; Deepwater Horizon oil spill and, 111n11, 115n22; endangered species and, 49–50, 114n21, 115n23; enframing and, 10–11, 24–25, 63, 99–100, 111n11; Hurricane Katrina and, 24–25; Indigenous knowledge and, 11

environmental movements, x–xi, 70

erosion control, 14, 29–31, 111nn1–2

Escobar, Arturo, 11, 104

experimentation: adaptive management and, 37–38, 113n14; dams and, 114n15; natural farming and, 59–69, 119n5, 120n15; performative, 97; science and, 37–38, 105, 113n14; skill and, 88–89; in water management, 37–38, 114n15–18

extended mind thesis, 107n1

Extinction Rebellion, 1

factory animal farming, 119n1, 119n4

falling height, 24, 99, 110n3, 131n8

The Falling Sky (Kopenawa), 91–96

farming: acting on in, 56–59; animal factory, 119n1, 119n4; chemicals and pesticides in, 57–59, 64–69, 121n20, 122n25, 122n29, 123n30; conventional, 56–59, 63–64, 119nn6–7, 121n18; dangers of conventional, 58, 63–64; desert revegetation and, 121n20; enframing in, 56–59, 64, 123n30; fertilizer and, 57, 64, 121n18; fires and, 84–86, 129n23; organic, 58–59, 119n7; plowing and, 57, 65; science and, 65–71, 84. *See also* natural farming

Fatherree, Ben, 111n12

fertilizer, 57, 64, 121n18

Feyerabend, Paul, 7

finding out, 4, 5, 7, 11–13, 37, 42, 44, 49–50, 61, 74, 89, 97–99, 102, 108n7, 114n15, 116n6

finished science, 98, 102, 108n7

fires: alpine, 127n13; appreciation of, 79; in Australia, 77–79; controlled burning and, 81–86, 125n1; danger of, 88; enframing and control of, 77–90; farming and, 84–86, 129n23; Gammage on, 80–90, 126nn6–7, 127nn9–12, 129n20; going back in time and, 89; Indigenous knowledge of, 15, 77–90, 126n5; Indigenous religion and, 87–88; management of, 77–90, 125nn1–2, 126nn3–7; Native Americans and, 79–80; nonhuman agency and, 83–84; pests, animals, and, 82–83; planning and surveillance of, 86; poiesis and, 84–90; Pyne on, 78–79, 127n13, 128n18, 129n23; science and, 84–86; technology and, 78–83

Fleck, Ludwik, 113n13

flooding, 110n9; animism and, 55; of Colorado River, 13, 34–35, 41–42, 112n3; Dutch water management and, 47–49, 116n4; in Grand Canyon, 4, 33–35; of Mississippi River, 14, 20–28

Foucault, Michel, 57

Franklin, Adrian, 78–79

frenziedness, 109n8

Fukuoka, Masanobu: cybernetics and, 69; natural farming and, 56–76, 120nn11–13, 121nn20–21; *The One-Straw Revolution* by, 58–76; religion and, 74, 123n34; science and, 66–69

Gammage, Bill, 77, 80–90, 126nn6–7, 127nn9–12, 129n20

Gates, Bill, 133n1

gestalt switches, 5, 13, 30–33, 51–52, 76, 94, 102

Ghosh, Amitav, 11, 23, 95

Gies, Erica, 117n8

Gilbert, Scott, 107n1

Glen Canyon Dam, 33–35, 40–44

God, 53–54, 117n10; *karma* and, 124n38

going back, 11, 53–54, 65, 89, 96, 121n21

graceless, 10

Grand Canyon, 4, 33–35

guardian spirits, 95–96

Hague, William, 7

Haraway, Donna, 107n1, 133n2

Haudricourt, André, 123n30

Heck cattle, 50–51

Heidegger, Martin, 32, 43–44, 47, 56, 109nn8–9; on enframing, 4, 10; on technology, 6, 99

knowledge and, 11, 15, 77–90; invasive species in, 14–16; letting go and, 6, 20–28, 49–52, 62–66; Mississippi River management and, 3, 14, 20–28; natural farming and, 9, 15, 56–76; nonmodern philosophies on, 12; religion and, 53–55, 117n10, 118nn11–12; spirits and, 15, 91–96, 128n19; as unknowable, 69–71

Needham, Joseph, 76

negative feedback devices, 114n20

neo-cybernetic paradigm, x

Netherlands, 46–55

New Orleans, Louisiana, 21–25

nomadism, 20, 25, 27, 46–47

nonduality, 132n10

nonhuman agency, 76; animism and, 91–95; in factory animal farming, 119n4; fires and, 83–84; foregrounding, 3–4, 10; in Indigenous knowledge, 9

nonmodern ontologies and philosophy, 12

not doing (*wu wei*), 75–76

Nyaminyami, 55

object-oriented philosophy, 107n1

oil drilling, 1, 111n11, 115n22

The One-Straw Revolution (Fukuoka), 58–76

ontology: action and, ix–x; Bateson on, xi; Christian, 53–55; cybernetics and, x, 39, 69; dualism and, xi, 2, 94–96, 98; Indigenous religion and, 87–88; knowledge and, 9; natural farming and, 69, 72–76, 123n30; nonmodern, 12; ontological turn and, 131n8; poiesis and, 9; of *shi*, 74–75; worldview and, ix

Oostvaardersplassen (OVP), 49–50, 116nn5–6

organic farming, 58–59, 119n7. *See also* natural farming

OVP (Oostvaardersplassen), 49–50, 116nn5–6

paradigm: emergence of, 5; Kuhn and, 100–101; neo-cybernetic, x; science and, 101–2; shift, from enframing, 49–55, 100–102

Pask, Gordon, 42, 70

pedagogy, x, 105–6

performance, 3, 30; in eel management, 17; poiesis and, 37–38, 98

performative epistemology, 7, 98

performative excess, 33, 99, 112n4, 113n9, 121n18

performative experimentation, 97

pesticides, 57–59, 64–69, 121n20, 122n25, 122n29, 123n30

physics, ix, 7, 17, 38–39, 93–94, 103, 108n6, 126n6, 131n8

Pickering Andrew: *The Cybernetic Brain* by, x–xi, 4, 6, 39–42, 70–71, 105; *The Mangle of Practice* by, ix, 3, 9

planetary consciousness, 109n9

plowing, 57, 65

poetic technology, 5–6, 115n27; AMP as, 35–37, 44. *See also* technology

poiesis: acting with and, 1, 15; Anthropocene and, 1; Colorado River and, 43–45; dance of agency and, 13–14, 105–6; in Dutch water management, 48–49; education and, 104–6; emergence in, 5; enframing and, 12–13; erosion control and, 29–31; fires and, 84–90; hybridity and, 12–13, 102; as iatrogenesis, 97; knowledge and, 31, 88–89; in Mississippi River management, 25–28; natural farming and, 65–77; ontology and, 9; paradigm shift to, 49–55, 117n8; performance and, 37–38, 98; poetic, 10, 61–76, 100; in practice, 14–15; revealing and, 4; science in, 7–9, 108n7; skill and, 88–89; as stance, 3–5; as technique, 5–6, 13–15, 99; term usage, 4–5; why we should care about, ix, 99–102; wicked problems and, 99; yin, yang and, 52, 96, 102, 123n33

politics, 7, 103–4

practical gestalts, 13, 102

Project Cybersyn, 41

Pyne, Stephen, 78–79, 127n13, 128n18, 129n23

"The Question Concerning Technology" (Heidegger), 99

race, 90

reciprocal vetoing, 41

religion: Buddhism and, 54, 71, 124n38; Christianity and, 53–55, 117n10, 118nn11–12; dualism and, 54–55, 72–76, 125n39; Eastern philosophy and, 72–76, 123n32, 123n34; enframing and, 53–55, 117n10, 118n11; Fukuoka and, 74, 123n34; Indigenous, 87–88; natural farming and, 72–76; social relations and, 89–90

resilience, ecological, 40, 129n23

resistance, 1–2

revealing, 4

revegetation of desert, 121n20

Rhine River, 32

Rice, Jim, 33, 112n7, 114n15

www.ingramcontent.com/pod-product-compliance
Lightning Source LLC
Chambersburg PA
CBHW030850270326
41928CB00008B/1308